T0146181

Aluminum Upcycled

JOHNS HOPKINS STUDIES IN THE HISTORY OF TECHNOLOGY
Merritt Roe Smith, *Series Editor*

Aluminum Upcycled

Sustainable Design in Historical Perspective

Carl A. Zimring

Johns Hopkins University Press
Baltimore

© 2017 Johns Hopkins University Press
All rights reserved. Published 2017
Printed in the United States of America on acid-free paper
9 8 7 6 5 4 3 2 1

Johns Hopkins University Press
2715 North Charles Street
Baltimore, Maryland 21218-4363
www.press.jhu.edu

Library of Congress Cataloging-in-Publication Data

Names: Zimring, Carl A., 1969– author.
Title: Aluminum upcycled : sustainable design in historical perspective /
    Carl A. Zimring.
Description: Baltimore : Johns Hopkins University Press, [2017] | Series:
    Johns Hopkins studies in the history of technology | Includes
    bibliographical references and index.
Identifiers: LCCN 2016019649| ISBN 9781421421865 (hardcover : acid-free
    paper) | ISBN 9781421421872 (electronic) | ISBN 1421421860 (hardcover :
    acid-free paper) | ISBN 1421421879 (electronic)
Subjects: LCSH: Aluminum—Recycling. | Metal products.
Classification: LCC TD812.5.A48 Z56 2017 | DDC 669/.7220286—dc23
    LC record available at https://lccn.loc.gov/2016019649

A catalog record for this book is available from the British Library.

*Special discounts are available for bulk purchases of this book. For more
information, please contact Special Sales at 410-516-6936 or
specialsales@press.jhu.edu.*

Johns Hopkins University Press uses environmentally friendly book
materials, including recycled text paper that is composed of at least
30 percent post-consumer waste, whenever possible.

For my parents

Contents

Acknowledgments

The inspirations for this book have been many. Teaching the rich interdisciplinary approaches to design at Pratt inspired this book, not least through conversations with my colleagues Uzma Rizvi, Frank Millero, Carolyn Shafer, Tetsu Ohara, Rebecca Welz, and Deb Johnson. I am especially thankful to Deb for bringing me into the Partnership for Academic Leadership in Sustainability, where I have been able to learn more about the state of sustainable design strategies from colleagues across North America in our annual meetings, Google hangouts, and a delightfully productive roundtable at the Association for Environmental Studies and Sciences meeting in New York City, which included Hélène Day Fraser, David Bergman, and Amy Deines. Thanks to the panelists and to fellow PALS Damien White and Kim Landsbergen from the liberal arts and sciences and Michele Jaquis from the design world.

Talking out themes for this book at interdisciplinary academic forums shaped the narrative, and I am especially indebted to Hanna Rose Shell for assembling a waste-themed panel that included me, Finn Arne Jørgensen, David Lucsko, and Roe Smith at the Society for the History of Technology conference in Portland, Maine. The participants at the Future of Waste workshop in Munich, including Michael Braungart, Kate Brown, Zsuzsa Gille, Sarah Hill, Roman Köster, Christof Mauch, Marty Melosi, and Song Tian, provided a rich discussion of both this research and our related interdisciplinary concerns in discard studies. Thanks also to Columbia University's Ecology and Culture seminar, especially Paige West and Nica Davidov. The Ecology and Culture seminar and a Pratt Institute Mellon grant also provided greatly appreciated financial support for the publication of this book.

In addition to the participants in those forums, several historians provided good feedback and advice. Special thanks to Nina Melechen for tracking down Buddy Miller's discussion of his guitars and Jim Longhurst for steering me right on bicycle history. My interdisciplinary colleagues in discard studies Steve Corey, Max Liboiron, Samantha MacBride, Tom McCarthy, Robin Nagle, and Kate

O'Neill and my mentor Joel Tarr continue to provide inspiration, which I hope is evident in these pages.

I had the pleasure to speak with many individuals who were designers or users of the goods covered in this book. Thanks to them all. Steve Albini, Tim Becker, Kevin Burkett, Hank Donovan, Tim Midyett, Conan Neutron, Chris Rasmussen, Jon San Paulo, Jay Tiller, and Nathan Zhang were helpful and patient with my questions; Jodi Shapiro, the founder of the first online Travis Bean inventory, provided invaluable photographs and responses to my interview questions. Emeco, Southside Guitars in Brooklyn, Classic Cycle in Bainbridge Island, Aston Martin, Ed Perlstein, Barney Taxel, the Eames Office, Touch and Go Records, and the band Tar were also generous with images and access to artifacts.

First and foremost among the archives and libraries that made this book possible is the Pratt Institute's library, including the extensive collections on design and an interlibrary loan staff that went above and beyond the call of duty in handling my numerous requests for rare items. Nick Dease, I owe you for all of those requests. Thanks as well to the staffs of the Henry Ford's Benson Ford Research Center, the National Archives, the Library of Congress, the Baker Library Historical Collections, Harvard University, the Metropolitan Museum of Art, the New York Public Library, and the Brooklyn Public Library.

Roe Smith's enthusiasm for the project put it on the path to Johns Hopkins University Press. At the press, Robert J. Brugger, Elizabeth Demers, Catherine Goldstead, Isla Hamilton-Short, Andre Barnett, and Kathryn Marguy provided guidance and assistance.

In addition to providing unconditional love and support, the crack architectural history guide and all-around ace in the hole Jen Potter gave me the bright idea to write a history of upcycling. Thanks always to Jen.

This book is dedicated to two people who are responsible for its genesis. My parents, Frank Zimring and Susan Hilty, exposed me to an appreciation of good design. I grew up in a household with Eames plywood chairs and Wassily leather-and-steel chairs. My parents also provided my first exposure to recycling on trips to take the many, many aluminum cans the household produced to a recycling drop-off center. Allowing a seven-year-old kid to crush cans one by one in a cool-looking machine had unintended consequences for my adult life. Those two aspects of my childhood come full circle with this book. Following my parents' example, I wrote this book for my students, whose ingenuity, optimism, and concern give me hope for the future.

Aluminum Upcycled

Toward a History of Upcycling

W aste is a product of design. Design that does not take into account the use, disposal, and potential reuse of the designed artifact generates waste materials that burden ecosystems and municipal waste management systems. Industry and schools of art and design in the twenty-first century recognize this, even as they produce a wider array of goods and materials. The history of industrial design shows the conveniences and comforts such production may bring; design is an important dimension shaping modern life.

What we drink is a good example. A soda is more than just a beverage. Whatever combination of carbonation, water, sweeteners, colors, and flavors that we consume in a matter of minutes is encased in a container intended to endure years of storage and transportation. Today, the material structuring the container in much of the industrial world is likely to be either aluminum (a can) or polyethylene terephthalate (PET) (a bottle). These materials are popular for their durability (since bottlers wish to minimize breakage and spillage) and lightness (reducing fuel expenses in shipping). These are modern materials that have inspired new styles, new consumption, and new waste since the second half of the twentieth century. As Jeffrey Meikle put it in his cultural history of plastic, these materials revolutionized modern life, expanding "beyond the purely material realm to the central meaning of a culture that is itself ever more malleable and inflationary."[1]

The advantages of these new, malleable materials include a world of goods unbound by past constraints of weight, inflexibility, and shape. The disadvantages include a host of ecological problems due to the excavation of raw materials and the growth of the volume and variety of discarded matter.

Today, recycling serves as a salve to those concerned with the consequences of post-consumer waste. In thousands of communities, consumers may place their cans and bottles in bins to be delivered to solid waste facilities (to be disposed of in landfills or incinerators) or in recycling bins. Since the 1970s, the public's perception of recycling is that it is an environmentally responsible, even moral practice to limit consumption's effects on the land, air, and water.[2]

The public's understanding of recycling obscures crucial dimensions of the practice. How recycling actually works only begins at the bin. For recycling to succeed, that bin must be collected by parties that can place the accumulated material with industries that will reprocess it and turn the discarded containers into new products.

Material scientists observe that recycling is not a simple act of turning old cans and bottles into new ones. A PET bottle is likely to have adhesive stickers, paper, and dyes in the paper and plastic. An aluminum can is likely to have a plastic coating and enamel or paint. The melting down of old containers requires separating this other material in ways that may produce toxins, which limits the desirability of the salvaged material.

For these reasons, practitioners in the scrap recycling industries have identified a practice of *downcycling*. Downcycling, unlike recycling, does not assume that the value of salvaged materials is static. Their material integrity or their economic (use) value may decline, and possibly both may happen at once. Analyses of collected materials reveal that many PET bottles in the recycling stream do not become new PET bottles; they are turned into other objects, such as plastic furniture.

The realities of downcycling are sobering reminders to advocates of recycling as part of zero waste strategies to reduce landfilling and incineration. They also raise the issue of extended producer responsibility. Would environmental degradation be reduced if, say, Coca-Cola did not use adhesives to affix stickers to its bottles or use paint on its aluminum cans? To what extent can products be designed to more successfully reclaim the use value of their component materials?

These questions raise the issue of how production choices may lead to the success or failure of material reclamation at the end of a product's life. In the twenty-first century, they have led to a focus on the converse of downcycling. Far from being downgraded, could those old cans and bottles be recycled into goods that are qualitatively superior to their previous incarnations?

While downcycling diverts (at least for a time) waste from landfills, sustainability advocates criticize the practice for being a suboptimal use of resources. In the 2002 *Cradle to Cradle: Remaking the Way We Make Things*, architect William McDonough and chemist Michael Braungart criticize downcycling for reducing the quality of a material over time and as a process that "can actually increase the contamination of the biosphere." Paints and coatings on metals from old cans and automobiles may produce dioxins and other toxic emissions that complicate material reuse as a closed-loop system of resource management.[3]

The term *upcycling* (popularized in fashion and industrial design since the late 1990s) reflects the creation of new goods from salvaged ones in a way that increases the value of the material. Designer Nathan Zhang of Beijing, China, for example, takes discarded blue jeans and other post-consumer fabrics, then works with a group of migrant women to turn the discards into shawls and capes. These designs are sold in Beijing, London, Paris, and New York City for several hundred dollars apiece. Since new blue jeans retail for approximately $5 in Beijing, the value of the used material is substantially enhanced by the design. With new jeans being so cheap, no market for secondhand jeans exists, so Zhang's designs transform used denim having no value into goods that far surpass that of new denim jeans.[4]

Zhang's work is representative of a new wave of zero waste fashion design (also known as ecouture fashion). A focus on how materials have been upcycled—rather than simply recycled—recognizes the changing valuation of the materials undergoing transformation.

Contemporary upcycling efforts range greatly in scale and kind. Artisan producers on Etsy tout their refashioning of old license plates into book covers as upcycling. Industrial designer Boris Bally's work includes chairs and plates fashioned from aluminum street signs. Bally does not remove the paint from the signs, so customers can identify the metal's previous use easily before they sit in the chair. Aspiring to be "the ultimate urban alchemist," Bally stated in 2014: "Making something people value from something they have discarded is the *ultimate* challenge. It's getting them to pay big bucks for your design made of their own discards."[5]

This work is a bridge between trained designers working in formal markets and the informal reuse of scrap materials in activities that increase the cultural and economic value of the items. This work goes on throughout the world, adding complexity to our understandings of the uses of modern materials. For example, artisans in West Africa convert scrap aluminum into cast aluminum cooking pots and utensils.[6]

Of late, however, the term *upcycling* has become associated with activities on a larger industrial scale. The allure of upcycling to attack environmental problems associated with waste has made its way from artisan producers like Zhang and Bally to large corporations. The giant shoe company adidas announced in 2015 a collaboration with designer Cyrill Gutsch's firm Parley to produce a sneaker with, as the press release declared, "a shoe upper made entirely of yarns and filaments reclaimed and recycled from ocean waste and illegal deep-sea gillnets."[7] Parley seeks to find design solutions for the problem of plastic ocean pollution. In speaking of these efforts, Ocean Revolution's founder, Wallace J. Nichols, argued: "Humans adapt. And one of the ways they're adapting is by turning this mess into other new products. They're doing science, they're doing research, they're communicating and they're being creative. That's what we do—that's what we humans do so well."[8]

Upcycling represents hope for responsible industrial production. The artisanal model of handmade goods differs in scale and process from the clothing company Patagonia's mass production of polar fleece from PET bottles or adidas's attempts to turn plastic found in the oceans into shoes. The material that Patagonia and adidas use is unrecognizable from its previous incarnation, lacking the shape and branding of its old body.

The industry trade group Keep America Beautiful (KAB) uses upcycling rhetoric in its advertising, showing a plastic bottle declaring "I want to be recycled" into various goods ranging from a hairbrush to a park bench.[9] Designers Norman Foster and Philippe Starck tout their use of secondary aluminum in chairs and tables as upcycling. And fashion designer Zhang identifies his use of discarded denim jeans to create capes that sell for $400 as upcycling.

These efforts add a new dimension to the perception of recycling as a form of environmental absolution for the waste-related sins of the modern world. Policies to encourage or mandate recycling are staples of zero waste strategies at the local, state, national, and international (in the case of the European Union) levels.

By the end of the twentieth century, municipalities in the United States hosted more than 10,000 curbside recycling programs with the rationale that collecting recyclable materials diverted them from sanitary landfills.[10] As recycling programs expanded, so did scrutiny of them. A 1996 article by John Tierney with the provocative title "Recycling Is Garbage" assailed collection programs as inefficient uses of money and energy and as a salve on consumerist guilt that did little to help the environment. Subsequent critiques of recycling included Susan Strasser's conclusion to her social history of trash in the United States in which she argues that curbside collection continues the "out of sight, out of mind" tendency to discard materials, which emerged during the industrial revolution.[11]

Strasser's analysis brings historical perspective to what many see as a behavior produced by the modern environmental movement. History has proved valuable for understanding what materials industries have sought to reuse at different points, what public and private systems have developed to reclaim materials, the evolving rationales for material reuse, and the technological innovations that have encouraged or aided recycling.[12]

An important continuity these histories share is the understanding that recycling exists because of the perceived value of the salvaged materials. Discards that are unwanted or uneconomical for manufacturers to use tend to have low recycling rates, even if public programs to collect the materials exist. For example, the US Environmental Protection Agency estimates the recycling rate for plastics in the United States at 8 percent of the material diverted from landfills despite widespread attempts to collect plastic bottles and shopping bags for recycling.[13]

In her book *Recycling Reconsidered*, New York City Department of Sanitation research director Samantha MacBride provided a nuanced critique of recycling as it relates to environmental health, noting that the collection and reuse of discarded materials too often is seen as the solution to a wide variety of environmental problems. Furthermore, the uncritical assumption that all consumer discards can be recycled puts burdens on collection programs, which struggle to manage items that were not designed to be easily disassembled, sorted, or reprocessed. The result, MacBride argued, is a solid waste policy and practice that "*isn't working* to reduce tonnage, toxicity, and continued growth of materials extractions and transformations in the United States or globally."[14]

MacBride's work is one of several twenty-first-century historical and journalistic studies that provide a better understanding of the ways in which recycling markets, technologies, and policies have developed. MacBride's analysis of how recycling does and does not contribute to reducing environmental burdens involves an important dimension of recycling often overlooked: the relationship between designers and materials reuse. The design of a product may encourage or discourage recycling. Historically, automobiles were difficult to disassemble, which left valuable steel unclaimed. While changes in recycling technology have allowed for the recycling of steel, the shredding that makes it possible creates toxic wastes due to hazardous materials in the designed vehicles, ranging from asbestos in brake pads to sodium azide in airbags. While the design of shredders allowed the harvesting of recyclable steel, the inattention to recyclability or disassembly in the design of the automobiles poisoned ecosystems and left processors subject to multimillion-dollar lawsuits and fines.[15]

Many industrial manufacturers now take an approach of designing for recycling, which takes into account the broader ecological consequences of landfilling, disassembling, or otherwise handling materials at the end of a product's life. Design for recycling is a facet in Apple's environmental strategy: the company replaced a mix of plastics and adhesives in 2008 with a unibody aluminum casing in large part because the former could not be easily recycled and the latter could. Aluminum can also be made elegant; Apple chief designer Jony Ive's team has created a series of sleek aluminum laptops. By doing this, Apple endorses McDonough and Braungart's conception of aluminum as a technical nutrient. In an Apple computer, the aluminum shell protects the electronics, and when the electronics are functionally obsolete, the shell can be recycled rather than landfilled. When Apple announced the new designs in 2008, CEO Steve Jobs called them "the industry's greenest notebooks."[16]

Apple's modifications are one example of how designers and design schools now attempt strategies to incorporate recycling and recyclability into products. Patagonia converts plastic bottles into polar fleeces, and the paper manufacturer Weyerhaeuser developed a branch focusing on reclaiming secondary fibers for use in its newsprint, paper, and corrugated containers. These industrial approaches reflect both economic incentives for salvage and design principles that value the environmental dimensions of manufactured goods. The influential industrial designer Dieter Rams argued in 1976 that design that does not take into account the "increasing and irreversible shortage of natural resources" is "thoughtless design," and he included environmental factors among his widely quoted ten principles of good design.[17]

Design schools now teach about recycling as part of sustainable strategies. These include developing process trees to analyze every stage of design and life-cycle assessments of goods (from the generation of primary materials to their construction, distribution, use, disposal, and reuse), using metrics to evaluate the environmental impact of designed goods, and developing tools such as the Okala strategy wheel to provide designers with helpful techniques to emphasize environmental factors when making design decisions. Institutional innovations to advance sustainable design include the consortium of art and design schools' Partnership for Academic Leadership in Sustainability (PALS) and Pratt Institute's Center for Sustainable Design Strategies.[18]

One popular approach in these developments is the championing of upcycling as a sustainable design strategy, and some discussion of exactly what is meant by "upcycling" is important, since applied definitions vary. Rubbish theory is useful. The sociologist Michael Thompson hypothesized that the value of goods and materials is dynamic, affected over time by social constructions that either reduce the perceived worth of decaying relics or inflate the worth of cherished antiques. Thompson argued that the rubbish category is a medium for the potential rediscovery of a past transient object and its subsequent reappearance as a durable good. One example is automobiles that decline in value until they are discarded as worthless, but several years or decades later are rediscovered and labeled as "antiques" or "classics." These classifications may be used to trace materials from new to discarded to reclaimed. Thompson's dynamic model of value has implications for the history of material use and reuse. To what extent is the recycling of materials an activity that changes the value of the recycled materials? To what extent is historical analysis of the dynamic value of materials useful in informing effective recycling practices in the present and future?[19]

The motives and logic of recycling activities past and present have been shaped by these social constructions. The word *recycling* (initially used in the petroleum industry in the 1920s to describe the recapture and filtration of petroleum for reuse) defines a process, but does not account for changes in the value of a material as it goes from product to salvaged material to new finished good. A Pepsi can refashioned into another Pepsi can may have a static value; that same can fashioned into part of a chair, automobile, or guitar creates something of far greater economic and cultural worth than the original can or collection of cans. Since salvage and recycling systems function based on the perceived worth of materials, investigating ways in which value has increased provides historical context for present-day attempts to reuse materials.

Constructing a History of Upcycling

How should this history be told? What method would be useful to understand the ways in which dynamics of value shape material reuse? One could assess the history of upcycling by looking at when the term gained popularity. This is a relatively recent phenomenon; the number of products on the e-commerce site Etsy tagged with the word "upcycled" increased from 7,900 in January 2010 to nearly 30,000 in January 2011 and 275,817 in September 2013.[20]

Scholars might analyze the development of publications using the term. That history is reasonably straightforward. A WorldCat search for the keyword *upcycling* in August 2013 produced 289 titles. Of these, most were from the previous five years. The term *upcycling* initially gained favor in Germany during the 1990s, with a handful of books on the subject published in German and English by 2003. But most of the books that describe upcycling have been published since 2007. These are largely design and crafts texts, with how-to instructions on reuse in specific applications, including Jason Thompson's *Playing with Books*, Bradley Quinn's *Textile Futures*, and Tristan Manco's *Raw + Material = Art*.[21] More general approaches to upcycling and zero waste principles may be found in Maggie Macnab's *Design by Nature* and Amy Korst's *The Zero-Waste Lifestyle*.[22]

The broadest assessment to date of upcycling as an organizing philosophy for industrial production was published by McDonough and Braungart, the creators of the Cradle to Cradle (C2C) certification program. They titled their 2013 book *The Upcycle: Beyond Sustainability, Designing for Abundance*, which was a manifesto for the future of industrial design. McDonough and Braungart described how industrial production should be reconceived to increase the value of existing materials with the ultimate goals to never cast any material from manufacture into sinks and never to create toxic wastes. For McDonough and Braungart, eliminating pollution is possible by ensuring all materials are either recyclable or biodegradable even as we produce goods at industrial scale. It is an ambitious goal, a use of design and material reuse that fosters abundance without the grave ecological problems found in industrial societies.[23]

Telling the history of upcycling through a list of publications reveals that the phrase has come into use relatively recently and is being seen with increasing regularity. If the goal is to state that the term is quickly gaining popularity in the twenty-first century, the method would be appropriate. It is a limited analysis, however. It obscures descriptions of practices that do not use the word and eliminates consideration of relevant practices that may be older than the term itself.

The Context of Industrial Ecology

If one takes "upcycling" to mean reducing waste while increasing the value of goods, the meaning of the term resonates with much of industrial ecology since the 1980s. Relevant examples include the work of the Rocky Mountain Institute in developing goods that produce rather than consume energy and other resources, the development of passive-house building principles, the Cradle to Cradle certification program and the 2002 book by McDonough and Braungart, and academic periodicals such as the *Journal of Industrial Ecology* and *Progress in Industrial Ecology*, which subject manufactured goods to life-cycle analyses to assess how to minimize the environmental effects of the goods' assembly, distribution, use, and disposal or reuse. This literature is broader, features approaches that are more systematic than many of the craft-oriented publications, and offers opportunities to see how a systems approach to the history of technology may reveal evolving environmental sensibilities in modes of production.[24]

This is a more engaging historical project than my first proposal since it delves into the values inherent in design over a longer period and with a more conscious philosophy. Industrial ecology in and of itself is old enough to assess the evolving patterns and methods in its history, and the broader approach to design in the environmental era is fruitful.[25] Such an approach touches on notions of natural capitalism as expressed in McDonough and Braungart's work and allows one to see how the values expressed by Rachel Carson, Barry Commoner, Paul Ehrlich, and other major voices of the modern environmental movement are or are not expressed by this design philosophy.[26]

But this effort, though more rigorous and useful than the first, still obscures important activity that should be considered in the history of upcycling. Much as the history of recycling begins further back in the past than the advent of curbside collection programs in the environmental era (and further back than the publication of Thompson's rubbish theory), the history of upcycling and downcycling should examine the methods and goals of manufacturers employing post-consumer and post-industrial materials throughout industrial history. This lens focuses on how and why industries have reused materials and allows a greater consideration of value and intention.

Histories of upcycling using this approach may lead to reappraisals of the automobile created from disused railways, the skyscraper built from the remains of demolished buildings, and even the mass-produced book assembled from rags. It may lead to philosophical debates on the values inherent in transforming plowshares into swords and swords into plowshares. I have chosen to discuss this history by focusing on a material largely employed since the mid-twentieth century, aluminum.[27]

Recyclable, Upcyclable Aluminum

Aluminum is a useful case study because it became a significant part of the waste stream in the middle of the twentieth century after mass production commenced during World War II; it is both abundant and versatile. Cambridge University engineering professors Julian M. Allwood and Jonathan M. Cullen published an analysis of sustainable materials in 2012, observing that steel and aluminum were particularly suited since they were both readily abundant and had the requisite material properties for wide applications in construction and industrial production. They concluded that "there aren't any other materials with such a good range of properties, available cheaply and in abundance."[28]

Recycling aluminum has both environmental and economic benefits when compared to the destructive consequences of producing virgin aluminum from bauxite. While mining and smelting bauxite into virgin aluminum requires vast amounts of energy and water, and has been blamed for ecological degradation in mining centers such as Jamaica, recycling involves 95 percent less energy than virgin production and far less toxins released into local air, land, and water.[29] Furthermore, the cultural history of aluminum's applications is rich and contradictory since the metal has been, as the historian Eric Schatzberg noted, both derided as ersatz and celebrated as modern.[30]

The history of secondary aluminum can provide perspective about contemporary upcycling. As a material that gained widespread adoption during World War II, the history of mass-consumed aluminum is relatively compact. The rapid inclusion of secondary aluminum in the total composition of aluminum (since the mid-twentieth century, even before recycling programs became commonplace, secondary material has made up more than half of the overall stream going into industrial production) means that the history of integrating salvaged material into new production is easily accessible. And the present-day prominence of aluminum in goods celebrated as upcycled provides a language to evaluate the historical record.

This is important because the term *upcycling* is becoming a staple of zero waste policies and design strategies. In an ideal world, successfully fashioning upcycled goods could provide economic incentives to divert materials from the waste stream and repurpose them (as McDonough and Braungart argued) as technical nutrients for a circular economy. Aluminum lends itself to reuse, which is evidenced by companies that use aluminum seeking recognition for upcycling. In 2013, McDonough and Braungart's C2C program reported certifying products from both the aluminum supplier Alcoa and the furniture manufacturer Herman Miller, and McDonough designed facilities for both companies.[31]

Aluminum's malleability has made it attractive to designers, who use it in both items of mass production and specialty products. Jony Ive's work to expand aluminum use at Apple reflects his interest in the metal. On a smaller scale, Ive worked with Marc Newson to design an aluminum desk and an aluminum camera to auction for charity. Far from the mass-produced laptops, the one-off desk was designed as a unique, covetable object, and it was auctioned in 2013 for $1.7 million.[32] Newson rose to prominence with an aluminum sofa in 1988. His welded Lockheed Lounge prototype was sufficiently iconic that it sold at auction 22 years after its construction for $2,098,500.[33] As Newson's and Ive's designs demonstrate, aluminum has a reputation in the design world as both elegant and sustainable.

But the reclamation of secondary aluminum to create durable goods of lasting value did not start in 1988 with Newson's aluminum lounge. How secondary aluminum was recycled in the past and how users treated and valued the refashioned products give us precedents for understanding how newly upcycled goods may affect the environment and economy in the future.

To an extent, such inquiry exists in companies seeking to understand the various costs and consequences of their products. Life-cycle assessments (LCAs) offer the potential of breaking down how material is sourced and what is required to assemble and transport the finished product to market. Life-cycle assessments may also provide an understanding of how (and for how long) users experience the product, and what happens (such as disposal or recycling) when the consumer has completed using it.

LCAs provide quantitative analyses of a product's embedded energy, toxicity, durability, and recyclability. Historical analyses add cultural perspectives to this framework. Why do some products that a designer intended to develop functional obsolescence experience great durability in their use and maintenance by the people who purchased them? (The history of aluminum-bodied automobiles may be informative.) Do the durability and value of otherwise quite stable goods change greatly over time, and why? (The history of aluminum-necked guitars offers some reasons.) Why might goods made out of secondary aluminum not become technical nutrients in an endless closed loop of industrial production? (The histories of these products and of aluminum furniture may help us understand the economic, environmental, and aesthetic complexities of material loops.)

This book is an attempt to understand the historical experience of upcycling aluminum and also the benefits and limitations that historical uses of secondary aluminum have offered. I hope this analysis contextualizes the contemporary practices termed *upcycling* and provides a more informed explanation of material reuse in industrial societies.

Chapter 1 focuses on the expansion of aluminum production during and after World War II. The vast resources that were expended to transform aluminum from a niche metal to a material of mass production were related to the US government's responses to the Great Depression and to the mobilization of military production. The reallocation of resources to scale up primary aluminum production during the 1940s had long-term economic and environmental consequences.

Chapter 2 explores the expansion of aluminum as a material of mass disposal. As industries incorporated aluminum into a myriad of uses, the metal found its way into the waste stream, compounding the environmental consequences discussed in chapter 1 with the burdens of disposal.

Chapter 3 describes the rise of secondary aluminum as a recyclable material. This was reflected in public campaigns to salvage aluminum, the private trade in secondary aluminum, and the ways in which salvaged aluminum became a commodity for industrial production. Manufacturers redefined aluminum production as a hybrid of primary and secondary materials, which were used for both disposable and durable goods.

The second half of the book consists of case studies of aluminum designs between 1945 and 2015. Chapter 4 focuses on the use of secondary aluminum in transportation, starting with new commercial designs of the airplanes that dominated wartime aluminum use and ending with Ford's environmentally based decision in 2014 to substitute aluminum for steel in the bodies of its most popular model, the F-150 pickup truck. I look both at designers' choices to use aluminum in vehicles and users' operation and valuing of the finished products.

Chapter 5 addresses the use of secondary aluminum in furniture designed for upper-middle-class to upper-class uses. One continuity in the long history of this use is the Herman Miller Company's production of such goods, ranging from the Aluminum Group furniture introduced by Charles and Ray Eames in 1958 to Philippe Starck's designs of the early twenty-first century. Herman Miller continues to produce both designs today in its McDonough-designed Greenhouse plant, and the company markets its products as environmentally responsible. While that language was absent from its 1960s catalogs, continuities in production exist. I also discuss how consumers have used and valued the furniture over time.

Chapter 6 looks at the use of aluminum in musical instruments, with particular focus on guitars. Between 1956 and 1985, several manufacturers (notably Wandré, Veleno, Travis Bean, and Kramer) produced aluminum-necked guitars for aesthetic and functional reasons. Stylistic obsolescence substantially reduced the value of these goods in the secondary market during the 1980s and 1990s, and new production ceased. However, shifting aesthetic tastes by the end of the century led to greater demand for these instruments, and in the twenty-first century new instruments by the Electrical Guitar Company and Obstructures have joined the now-vintage models in selling for thousands of dollars apiece. How these instruments have been used and why their value has changed reflects the complexities in the experience of upcycled goods.

The book concludes with some lessons learned from the history of upcycling, how they may be applied to an understanding of current aluminum reuse, and how they may be instructive for approaching the reuse of other upcycled materials, such as plastics.

To understand how aluminum came to be upcycled, one must understand how and when aluminum became a material of mass production and a material associated with sustainable production. The history of this most modern metal is intertwined with the economic decisions made across the globe during World War II.

PART I / Creating a Technical Nutrient

Chapter One

From Scarcity to Abundance

O n Elm Street is a building that behaves like a tree. Most of the architecture on this street in Oberlin, Ohio, dates from the late nineteenth century;
brick Victorians and wood-framed houses with expansive lawns and gardens
dominate the block. But the Adam J. Lewis Center for Environmental Studies at
Oberlin College was built at the end of the twentieth century to do more than
look picturesque.

It is a striking building, featuring an atrium of glass and aluminum that wel-
comes visitors and provides natural light during the day. The large windows are
evident in the building's seminar rooms, allowing environmental studies faculty
to demonstrate to students how design and resource use are linked. The building's restrooms do not connect to a septic tank or the local sewer system. Instead,
wastewater stays within the facility, traveling to a "living machine" that uses
plants to treat the sewage. The arched roof and adjoining parking lot feature pho-
tovoltaic panels that generate energy for the building. The wood in the building
was sourced from local forests; many of the other materials used in construction,
including the aluminum visible in the windows and doorframes, was recycled.
These features were all included by design, emanating out of discussions between
an environmental studies professor, David Orr, and an architect, William Mc-
Donough, about, as McDonough put it in *Cradle to Cradle*, "the idea for a building
and its site modeled on the way a tree works."[1]

Two blocks east and one block north of this revolutionary structure lies a more

The Adam J. Lewis Center for Environmental Studies at Oberlin College, with exterior clad in aluminum. Courtesy Barney Taxel

conventional house, but 64 East College Street has a history that made the Adam J. Lewis Center for Environmental Studies possible. This is the house where Charles Martin Hall grew up, and where he and his sister Julia conducted chemistry experiments in the woodshed in back. The Hall siblings had been students in Frank Fanning Jewett's class when the professor remarked in 1880 that (as Jewett recalled years later) "if anyone should invent a process by which aluminum could be produced on a commercial scale, not only would he be a great benefactor to the world but would also be able to lay up for himself a great fortune." Jewett remembered Charles telling a classmate, "I'm going for that metal."[2]

Charles Hall spent the next six years trying to produce the metal. In the winter of 1886, Charles and Julia (the extent of her involvement is the subject of historical speculation) discovered the electrolytic process for synthesizing aluminum from bauxite.[3] The method, discovered at roughly the same time by the Frenchman Paul Héroult, became known as the Hall-Héroult process, and it allowed production of aluminum at a scale sufficient for industrial use. Charles Martin

Hall became a multimillionaire from this work, and he put some of this wealth into Oberlin College before dying in 1914. His benevolence contributed to the college's financial stability and its eventual ability to construct the Lewis Center. Visitors to the Hall family's modest house can learn about its central place in industrial history through tours and the plaque commemorating Hall's achievement.[4]

The historical path from the woodshed at 64 East College Street to McDonough's design for sustainability is much longer than the ten-minute walk in Oberlin. Hall moved east and founded the Pittsburgh Reduction Company with financier Alfred E. Hunt in 1888 to take advantage of his discovery. Their firm prospered, becoming the Aluminum Company of America, now known as Alcoa. Pittsburgh bears evidence of this firm's success today, with an aluminum and glass library on Carnegie Mellon University's campus named after Hunt and an aluminum-clad thirty-story skyscraper in downtown now known as the Regional Enterprise Tower. This structure used to be known as the Alcoa Building, and at the time of its construction in 1953 *Popular Mechanics* called it "the lightest building of its size in the world."[5]

Hall became wealthy from his Pittsburgh-based company, but the path to that wealth stretches further, across the planet to the mines that produced the raw materials to fashion aluminum. Over the twentieth century, aluminum producers, including Alcoa, used the Hall-Héroult process to create modern materials at a resounding ecological cost. The money and materials from this work went into the Adam J. Lewis Center, and the history of primary aluminum production is important for evaluating McDonough's material choices for this example of sustainable architecture.

Aluminum in the Nineteenth Century

Although aluminum is the most abundant metal found in the earth's crust, it is not found freely in nature, and before the Hall-Héroult process was invented, extracting the metal from bauxite or compounds was difficult. Evidence of 7,000-year-old clay pots containing aluminum silicates has been found in Persia, but the name *aluminum* dates to British chemist Humphry Davy's isolation of the metal in 1808. Beginning with Danish chemist Hans Christian Ørsted's 1825 experiments, nineteenth-century scientists worked to produce aluminum from bauxite.

Aluminum is most commonly found in the earth's crust as an oxide within bauxite, which contains about one-third aluminum oxide (also known as alumina). Aluminum atoms have a stronger attraction to oxygen than do iron or carbon, so separating aluminum from its oxides proved difficult until 1887. That year, Karl Joseph Bayer discovered that if bauxite is washed in caustic soda (sodium hydroxide, NaOH), the alumina within it dissolves. After the solution is filtered, dried, and heated to 1,050°C, the alumina is released as a white powder. With the advent of the Hall-Héroult process and the Bayer process in the 1880s, the techniques for creating primary aluminum were in place. The Bayer method purifies mined ore so it is fit for smelting in the Hall-Héroult process. The Bayer process leaves a strongly alkaline waste in the caustic, material that today is commonly known as "red mud."[6]

The Hall-Héroult process helped to reduce the price of aluminum from $4.86 per pound to 78 cents per pound between 1888 and 1893, and aluminum became more common in artisan crafts and tableware by the turn of the twentieth century. The Pittsburgh Reduction Company built an American industrial giant based on the process; in Western Europe, Héroult licensed his patent to the Swiss company Aluminium Industrie Aktien Gesellschaft (AIAG), subsequently part of the Alusuisse-Lonza group.[7] The Swiss concern was one of two large European aluminum companies during the early twentieth century. In France, Henri Merle's Compagnie des Produits Chimiques Henri Merle had a state-granted monopoly on aluminum production; the company morphed into the Société des Produits Chimiques d'Alais et de la Camargue in 1897 and, after mergers, became officially known as Pechiney in 1950.[8]

Innovation allowed designers to create new goods from this light, durable material. Aluminum was valued in the nineteenth century for its durability and lightness, yet even with the Hall-Héroult process, the energy requirements of smelting aluminum from bauxite limited use of the metal before World War II.

Flying Machines

That said, aluminum made inroads into the new field of aviation in the early twentieth century. An airplane made of a light material more durable than wood or canvas might be safer to fly, which would be doubly important if the airplane were used in combat operations. A key advance following the Hall-Héroult process was the innovation by Alfred Wilm of the Metallurgical Department at the German Center for Scientific Research to add magnesium and manganese to aluminum in 1906. The resulting alloy took two days to harden, but when it did, it became three times as hard as when the initial alloy was mixed.[9]

Wilm named his alloy *duralumin* and put it into commercial production in

1909. Most of the 13 tons produced in 1910 were purchased by the British Vickers Company, which used 10 tons to build a dirigible called the *Mayfly*. The aircraft broke in two during preparations for its first flight, stigmatizing the German alloy in the eyes of British industrialists. However, duralumin would soon become the backbone of German air technology, including zeppelins and the early airplanes used in World War I.[10]

The outbreak of war in Europe in 1914 spurred engineers to work on aluminum aircraft. Hugo Junkers developed the first all-metal airplane in Germany. Junkers was an engineering professor in Aachen who had worked on a failed prototype for a plane with aluminum wings after he quit teaching and built a laboratory.[11] The iron-clad J-1 was tested in December 1916. The weight of the iron caused the J-1 to be too heavy and underpowered to be effective. Junkers then experimented with the aluminum alloy called duralumin. Manufactured by Dürener Metallwerke, duralumin was as strong as steel at one-third the weight. This metal proved suitable for Junkers's J-4, a biplane designed for low-level ground attacks. Although the J-4 was slower than lighter wood-and-fabric-winged airplanes, its all-metal body was more resistant to artillery fire from ground troops, and the German military used it with great success. Other industrialized countries worked to develop similar alloys; Alcoa marketed its 17S alloy to airplane manufacturers in the United States.[12]

Duralumin was developed with military applications in mind. The metal was not durable, corroding into a white powder when exposed to oxygen. This could be a problem in flight, since salty sea air accelerated the process, causing the metal to fail while at top speed. Aviation historian Tom Crouch argued that this weakness was not a crucial problem for military purposes because "the life expectancy of a warplane was short anyway, and industry would get by for another few years before it solved the problem."[13]

In addition to Junkers's biplane, German uses of duralumin included the Zeppelin Company's large flying boats with externally braced wings. Zeppelin designer Claudius Dornier started his own firm in 1918 and created the D-1 fighter plane with a duralumin structure.[14]

Military necessity had spurred innovation that might have benefits beyond the battlefield. But the needs of combat aircraft differed from civilian needs. Military airplanes had to withstand artillery fire to complete individual missions; civilian airplanes had to be durable over years of transportation. The alloys used during the war were subject to corrosion, not a problem if the intended use was for weeks or months, but not suitable for the demands of civilian applications.

A major advance by Alcoa researchers came in 1927 with the development of alclad, described by Crouch as "a very thin layer of soft but corrosion-resistant pure aluminum [that] was bonded to either side of a sheet of duralumin." Alclad proved very successful; by the late 1920s standard sheets of wrought aluminum alloys (17S for the Boeing 247; 24ST in Douglas aircraft) were treated with the alclad process and used in a series of new aluminum-bodied American aircraft.[15]

The designer Jack Northrop left Lockheed in 1928 and developed the Alpha for the Avion Corporation. The Alpha, according to Crouch, was "a sleek, low-winged, all-metal monoplane," which helped trigger the design revolution for all-metal aircraft.[16] Impressed by the Alpha, Boeing acquired Avion. Boeing's first all-metal plane was the model 200 Monomail in 1930, followed by the B-9 in 1931, which was the first operational military aircraft made of metal.[17]

Several other aluminum designs for commercial aircraft followed in the 1930s. United Airlines purchased the Boeing 247.[18] Douglas developed the DC-1, which first flew in 1933 as an aluminum commuter plane for TWA. Durable aluminum meant that manufacturers could create larger aircraft to carry passengers or cargo in light structures that would not break up in flight. The planes were expensive, but Boeing, Douglas, and Hughes could afford to make them.[19]

Modernity and Large Technological Systems

Designers and the aviation industry desired aluminum from the end of the nineteenth century, but the rise of aluminum from scarcity to abundance is not a story of a free market developing to fill demand. Indeed, the US government's successful antitrust case against Alcoa provides evidence that the market was not free in the late 1930s. Thomas P. Hughes characterized the Manhattan Project and its influence on military missile, air defense, and communications projects as the products of large technological systems, developed as collaborations among the government, industries, and universities.[20]

Governments in the Americas, Europe, and Asia fueled the global expansion of aluminum manufacturing during the 1930s and 1940s. They built the capacity for production in conjunction with industries, developing energy regimes that finally allowed for the mass conversion of bauxite to aluminum. They maintained and protected supply chains to bring bauxite to smelters and then new aluminum to factories. And they acted as consumers of aluminum in the form of military aircraft. This mobilization revolutionized aluminum production and distribution during World War II and shaped the metal's postwar history. The global production and consumption of aluminum in the early twenty-first century rely on the decisions made in the 1930s and 1940s.

American aircraft companies could afford to make aluminum planes in the 1930s for the same reason that the Soviet Union's manufacturers and Germany's Junkers could; their governments invested in the large technological systems that made mass production of aluminum possible. The major hurdle to building aircraft from aluminum was the high cost of the metal due to its intensive energy demands. Aluminum smelting, in Mimi Sheller's words, is one of the most energy-intensive production processes on earth. "Smelting uses an electrolytic process in which a high current is passed through dissolved alumina in order to split the aluminum from its chemical bond with oxygen. The electrochemical smelting of aluminum from refined bauxite ore requires between 13,500 and 17,000 kWh of electricity per ton, more energy than any other kind of metal processing."[21]

The development of hydroelectric dams in the American West during the New Deal and the demand for light aircraft during World War II made mass production of the metal both possible and desirable despite the environmental costs. After exceeding 100,000 metric tons produced worldwide for the first time in 1916, global aluminum production hit a high of 280,000 metric tons in 1929 before declining during the Great Depression. Global production then hit a nadir of 142,000 metric tons in 1933 before gradually rising over the rest of the decade.[22]

Major hydroelectric dam projects provided the vast amounts of energy required to smelt virgin aluminum from bauxite; American hydroelectric capacity gave the United States a distinct advantage over Germany in the production of new aluminum after the start of World War II. The Bonneville Dam on the Columbia River in the Pacific Northwest produced energy for a burgeoning aluminum industry (so much so that the giant concern Anaconda Copper moved into aluminum production in 1955 to take advantage of the abundant power), and the Bonneville Power Administration established energy agreements with new production facilities in the region. In 1940, Alcoa opened a gigantic smelter on 218 acres in Vancouver, Washington, across the river from Oregon.[23]

Felix Padel and Samarendra Das argued that global "histories of aluminum and dam construction go hand in glove, linked from birth."[24] Dams and aluminum production after World War II expanded to the Suriname River and tributaries of the Amazon River in South America, the Zambezi River in Mozambique, and the Yangtze River in China.

In the United States, dam projects were popular during the New Deal because of the jobs they produced. Once complete, dams provided the capacity to create aluminum at an unprecedented scale. By 1944, the Bonneville Power Administration was producing 9.5 billion kilowatt hours annually, making it the nation's third largest electrical power system. The Pacific Northwest produced more than 315,000 tons of aluminum ingot through 1948, a year in which it was responsible for 47 percent of domestic aluminum production.[25]

The series of technological innovations in aviation since World War I meant that Germany, the United Kingdom, Russia, Japan, and the United States all had access to aluminum-bodied fighters and bombers by the time Germany invaded Poland in 1939. The scale of conflict over the next six years required production of those airplanes on an unprecedented scale, as the Allied and Axis powers fought air wars across the globe.[26]

American military propaganda explicitly linked hydroelectric dams with the ability to make bombers from aluminum. One poster championed "Kilowatts to Kill the Rats! TVA Power Gives Us Aluminum for Planes," with depictions of angry rats with symbols of the Axis nations of Germany, Japan, and Italy looking up at bombers attacking them. Looming behind them all was a hydroelectric dam.[27]

In 1945, the economic analyst Charlotte Muller concluded that Alcoa "has been forced into a dominant position in the international hydroelectric power industry by its dependence upon cheap power."[28] Muller identified Alcoa's acquisition of site rights in Niagara Falls, Quebec, the Tennessee Valley, and Norway between 1905 and World War II as evidence of this dependence, noting that the expansion of capacity by the Tennessee Valley Authority in the late 1930s and early 1940s was a boon to the growing aluminum concern. "As the pressure of wartime power needs impelled the TVA toward an expanded construction program, a contract was finally signed with Alcoa on August 14, 1941, turning over the Fontana site to the TVA. No cash was paid, but Alcoa was compensated by the better storage facilities which brought a more even river flow to Alcoa's downstream dams at Cheoah and Calderwood."[29] Muller identified domestic hydroelectric dam projects relevant to Alcoa in Tennessee, Oklahoma, and North Carolina, and especially the gigantic projects in the Pacific Northwest. The Bonneville and Grand Coulee projects on the Columbia River combined with the TVA dams to create "a tremendous power potential beyond the requirements of the aluminum industry and beyond the ordinary investment capacity of private capital."[30]

"Kilowatts to Kill the Rats! TVA Power Gives Us Aluminum for Planes." US World War II propaganda poster celebrating hydroelectric capacity to produce military aluminum. Courtesy National Archives

Mass Production of Military Aircraft

With the energy requirements met, public and private aluminum operations required one more ingredient for mass production: bauxite. Getting this material involved exploiting resources under domestic control during wartime, as well as establishing safe routes to transport the material over growing distances. An analysis conducted one year after World War II ended concluded that "almost half the known bauxite resources in the world" were located in Europe, with the largest reserves in Hungary, about half as much located in Yugoslavia, and smaller but significant amounts in France, Greece, Romania, Italy, the Soviet Union, and the Norwegian-controlled Arctic north.[31] France was the center of bauxite mining, especially in its Var and Herault districts near the Mediterranean coast. French bauxite in the mid- to late 1930s fed the domestic aluminum industry and supplied the United Kingdom and (until France imposed restrictions over military developments in 1935) Germany. Most Hungarian bauxite mined was exported to Germany.[32]

By 1938, Germany was the world's largest producer of aluminum, material that produced a large and modern military air force. "Germany alone accounted for thirty percent of the 1937 increase in global aluminum consumption," industry historian George David Smith wrote, as the Third Reich, violating the Versailles Treaty's prohibition on rebuilding a German air force, created the Luftwaffe. "By the end of 1937, Great Britain began to signal . . . that it would need more aluminum to expand the Royal Air Force."[33]

Germany's four-year *Vierjahresplan* for war preparation in 1936 made its dominance in aluminum possible, with primary production from AIAG, Aluminiumwerk Bitterfeld, and Vereinigte Aluminium Werke (the latter was state-owned). In 1941, Hermann Göring led a reorganization of aluminum production involving bauxite mining and smelting facilities throughout much of the European territory Germany occupied during the war.[34]

Soviet aluminum production had begun with the opening of a smelter in Volkhov in 1932. Threats from the German invasion led to production relocating during World War II to the Urals and Siberia. The expanse of the Soviet Union allowed the entire vertical chain of production to take place within its borders, including vast extraction of bauxite between the 1930s and 1980s from the Urals, Central Asia, Turgay, Kazakhstan, Yenisey, Timan, Altai-Sayan, Tikhvin-Onega, Voronezh, and Ukraine. In the two decades after World War II, Soviet smelting of aluminum expanded into new facilities at Kandalaksha, Nadvoitsy, Volgograd, and several Siberian cities following completion of a series of hydroelectric dams there.[35]

Operations in the Americas included large bauxite mines in Jamaica and Arkansas, with expanded hydroelectric production in the Tennessee Valley, along the Columbia River, and in Canada. The Aluminum Company of Canada boasted in 1952 of its $380 million plans to expand hydroelectric power in Quebec and British Columbia and bauxite mining in Jamaica, expansions that built on capacity developed during World War II.[36]

Germany's expansion between 1936 and 1940 brought reactions from the Soviet Union, the United States, France, and the United Kingdom. Soviet aluminum production expanded. The American, French, and British militaries increased their orders to industry, producing a spike in global production even before Germany invaded Poland in September 1939.[37] Military investments in aluminum for aviation helped the rise at the end of the decade, with global production increasing to 482,000 metric tons in 1938 and 787,000 metric tons in 1940, exceeding a million metric tons the following year, and then almost doubling (to 1.95 million metric tons) in 1943.[38]

This expansion required a significant marshaling of resources and new organizational efforts. The US Military Aircraft Program debuted in late 1940, and in response, Alcoa began construction on a sheet mill at Alcoa, Tennessee, with the capacity of producing 5 million pounds of aluminum a month. Though huge, the new facility quickly proved too small to meet demand.[39]

The historian George David Smith described Alcoa as "being pilloried" for inadequate production in early 1941. The Office of Price Management (OPM) criticized the company for being unable to meet demand given the problems of "increased working inventories of aircraft plants, increased military requirements for small aluminum items, hoarding of secondary aluminum [alleged by OPM, a common criticism it made of secondary dealers], a fall in Canadian imports, and bottlenecks at the finishing end."[40]

Existing tensions with Alcoa's civilian customers were exacerbated by the war. R. J. Reynolds and Alcoa clashed in 1939, spurring important developments. Alcoa cut Reynolds off from its supply of ingot, and Reynolds (which was making aluminum packaging, powder, and paste) sought alternative sources of the metal. The company's president, Richard S. Reynolds, knowing that Germany and Switzerland had developed abundant supplies of aluminum, traveled to Central Europe, where he realized how extensively Germany had developed its military aviation program. Reynolds returned to the United States and urged Alcoa's A. V. Davis to triple its aluminum manufacturing for aircraft production.[41]

According to Reynolds, Alcoa was unresponsive to the German threat, and he went to the Reconstruction Finance Corporation seeking economic aid to develop competing aluminum production. Convinced, the RFC granted Reynolds loans secured by a mortgage on his eighteen plants to construct a smelter in Washington state and a smelter and sheet mill in Listerhill, Alabama.[42]

Alcoa's monopoly was further undermined by the US Senate's Truman Committee, which urged the federal government to begin producing aluminum, which it did beginning in 1941. Demand from the US military led to a 600 percent increase in domestic aluminum manufacturing between 1939 and 1943.[43] Virtually all of this production came from Alcoa, which was responsible for more than 90 percent of domestic aluminum production in 1944.[44] However, in the words of Smith, the US government's intervention into aluminum production "effectively end[ed] Alcoa's monopoly on the domestic market," which "set in motion both the mass production of the metal and the context for its growing accessibility and affordability."[45]

The federal government had been concerned about Alcoa's monopoly of aluminum before the war; in 1937, the Department of Justice had initiated the federal antitrust suit *United States v. Alcoa*.[46] In March 1945, the Second Circuit Court justice Learned Hand found that Alcoa had maintained an illegal monopoly of the aluminum ingot trade. The antitrust action opened up the market, allowing Reynolds and Henry J. Kaiser to build on their wartime expansions and become viable competitors to Alcoa.[47]

Kaiser Shipyards acquired aluminum production facilities from the federal Surplus Property Board. Reynolds had used aluminum foil in prewar cigarette packaging. It expanded its operations and established independent aluminum fabricating works to produce a wide variety of aluminum products beyond the packaging and powder it had made with the use of Alcoa ingot in the 1930s. The Surplus Property Board sold most of the government-managed aluminum plants to non-Alcoa interests, allowing Kaiser and Reynolds to compete with the giant.[48]

Manufacturers rapidly scaled up production of aluminum-bodied airplanes, reshaping air combat. Tom Crouch concluded that "the pressures of war pushed existing technology to its limits," leading to several new designs that became dominant in the American military. "Airplanes introduced during the war, including the P-51, F4U, and B-29, remained in active service into the next decade."[49]

Crouch described each of these airplanes in *Wings: A History of Aviation from Kites to the Space Age*. One example of what aluminum meant was the development of the Boeing B-29. The plane, an aluminum-bodied "superfortress," became the staple bomber in the US military; their relative lightness and power allowed these planes to travel high and carry heavy bombs. B-29s were used in operations that included the killing of more than 20,000 residents of Dresden, Germany, by 3,900 tons of bombs and the immediate killing of more than 60,000 residents of Hiroshima, Japan, by the first atomic bomb ever dropped as an act of war. The B-29 that dropped that bomb, the *Enola Gay*, is now on display at the National Air and Space Museum in Washington, DC; with its polished aluminum finish, it is one of the most famous artifacts in aviation history.[50]

Soviet attempts to engineer a duplicate of the B-29 resulted in the Tupolev Tu-4. North American Aviation's P-51 Mustang, with a fuselage constructed entirely of aluminum, became the dominant American fighter. Clad with pantal (a German aluminum alloy incorporating titanium), Junkers's Ju-87 Stuka became the workhorse of the eastern front, usually fighting the Soviets' heavy steel-clad Ilyushin IL-2. By 1945, Alcoa, AIAG, Reynolds, Kaiser, Sumitomo, and the Soviet Union had produced vast amounts of aluminum to fight World War II. The metal was no longer scarce.[51]

Mass Production on a Global Scale

While the end of Alcoa's monopoly spurred competition in the aluminum industry, it cannot explain the vast expansion of aluminum production during and after World War II both within the United States and worldwide. The large technological systems built by national governments starting in the 1930s had been necessary for the war. Remaining in place after the war, dams fueled the continued production of primary aluminum in North America, Europe, and Asia.

The year 1943 marked the high point for global production until the middle of the Korean War, with 2.06 million metric tons produced that year. The global price of aluminum in 1950 declined to $1 per pound in constant 2000 dollars from $8 per pound in 1910. The alloys used in wartime aviation (notably 7075-T6, introduced in 1943) improved durability and enhanced the value of the metal in practical applications even as the cost of the material declined.[52]

The federal government, wary of potential monopolies, intervened and opened competition to R. J. Reynolds's Metals Division and Henry J. Kaiser's Metals Division. Concurrently, Germany and the Soviet Union smelted aluminum for aviation. By the end of the war, aluminum was an important element of mass production in the United States and Europe, and for the first time, large quantities of used aluminum were discarded.

Aluminum retained its role in military production. Dewey Anderson, the author of *Aluminum for Defence and Prosperity*, wrote in 1951 that aluminum had become the "most important single bulk material of modern warfare," adding that "no war can be carried to a successful conclusion today without using and destroying vast quantities of aluminum."[53] The rapid development of American air power relied on aluminum. In warplanes, 90 percent of the wings and fuselage, 60 percent of the engine, all of the propeller, and various wires, rods, mechanical systems, and implements were composed of aluminum.[54]

The aluminum industry recognized, however, that peacetime military production would not be sufficient to use the much larger quantities of the metal that individual firms were now able to produce. In the United Kingdom, postwar efforts to champion innovation included the "Britain Can Make It" industrial exhibition held at London's Victoria and Albert Museum in 1946. This expo included more than 5,000 prototypes and new products from more than 1,000 British firms, including aluminum products that are discussed in chapters 4 and 5.[55]

The American producers hired designers to work either in-house or with potential customers to create new goods from aluminum. Reynolds shifted in-house designer Jim Birnie to head the company's Styling and Design Department in 1950. Five years later, Alcoa created a Market Development Department and put it under the direction of Fritz Close. Close subsequently hired the industrial designer Sam Fahnestock. In 1956, Kaiser Aluminum hired Franklin Q. Hershey away from the Ford Motor Company to lead its Industrial Design Department.[56]

These designers allowed aluminum manufacturers to expand the market for their product beyond aviation and into many facets of the postwar economy. Education about ways in which aluminum might be employed in new designs would expand demand, and thus education was an important dimensions to aluminum manufacturers' strategies in the 1950s.[57] Design historian Dennis P. Doordan argued that this education via industry publications and other communications worked to stimulate a culture of creative engagement between the companies and designers. Not only did the manufacturers discuss the strength and malleability of the material, they also emphasized aluminum's utility in artistic expression and affordability in mass production, blending the concerns of art and commerce.[58]

Alcoa's corporate headquarters in Pittsburgh was a symbol of this effort. Alcoa contracted the architects Wallace Harrison and Max Abramovitz in 1950 to develop a 30-story skyscraper that would be a functioning advertisement for the company's wares. Completed in 1953, the Alcoa Building featured a skin of 1/8-inch-thick Alumilite panels that were riveted into place instead of welded. Inside the building, aluminum was the raw material for the wiring, the plumbing, the cooling tower, and the finishes of surfaces of drinking fountains and elevator doors. At the time of its construction, the Alcoa Building was hailed as the lightest skyscraper in the world. Although Alcoa vacated the building in 2001, it remains in use by other businesses, a conspicuous example of aluminum architecture in downtown Pittsburgh.[59]

The Alcoa Building was the most ambitious use of aluminum in architecture, but other architects embraced the possibilities of the now-abundant metal, adding it to designs throughout the 1950s. The Alcoa Building was surpassed in 1957 by 666 Fifth Avenue in New York City as the largest skyscraper with exterior aluminum panels. By the time of that building's completion, the father-and-son team of Eliel Saarinen and Eero Saarinen had designed a massive aluminum and glass 24-story building for General Motors' Technical Center in Warren, Michigan. Opened in 1956, the project was called by its client a place "where today meets tomorrow." Eero Saarinen enjoyed working with aluminum and incorporated it into several projects before his death in 1961; his work on Dulles

International Airport figures into one of the design innovations discussed in chapter 5.[60]

The products Alcoa featured in its headquarters thrived in the marketplace. Following the Korean War, production of aluminum accelerated, growing each year through 1974. Global production hit 3 million metric tons in 1955 and 4 million metric tons in 1959. The contemporary observer Alfred Cowles noted that "aluminum in the latter part of 1957 may have overtaken copper as the second most important metal in the world [after iron] on the basis both of tonnage and value of metals consumed by industry."[61]

The now-competitive aluminum market produced material beyond demand for several years. The 1957–1958 recession sharply reduced demand at a time when Alcoa, Kaiser, Reynolds, and smaller producers had greater parity and cause to undercut each other. Alcoa corporate historians Margaret B. Graham and Bettye H. Pruitt noted in their study that American production of aluminum had increased by a factor of 10 since 1939, and in 1957 Reynolds, Kaiser, and Anaconda controlled 55 percent of the American market. A recession in 1957–1958 "burst the bubble of growth and profitability that had enveloped the industry since the war," leading to falling prices and profits through 1961 and then a slow recovery. By 1965, profits were still 36 percent below the high of 1956.[62]

Despite economic problems in the United States, the global growth of aluminum production continued unabated through the 1960s. As it did, the dominance of the world's largest aluminum producers (including Alcoa, Reynolds, and Kaiser in the United States; Alcoa's spin-off Alcan in Canada; and Alusuisse and Pechiney in Europe) gradually diminished, accounting for 84 percent of the global market in 1955, 72 percent in 1965, and less than 58 percent in 1981.[63]

Share diminished not because the individual companies were less prolific, but because competition spurred overall growth. Global aluminum output tripled in less than 20 years, exceeding 5 million metric tons in 1962, 6 million metric tons in 1965, 7 million metric tons in 1967, 8 million metric tons in 1968, 9 million metric tons in 1970, and 10 million metric tons in 1971. It peaked at 13.2 million metric tons in 1974 before the global energy crisis reduced production to 12.1 million metric tons in 1975. Global production recovered to exceed 15 million metric tons in 1980. Aluminum had become a material of mass production.[64]

Ecological Disaster

The environmental and economic consequences of aluminum's mass production since the mobilization for World War II were dramatic. The energy required to create primary aluminum came with a price. Hydroelectric dams disrupted aquatic ecosystems and displaced indigenous peoples. Coal-fired electric plants required the mining and burning of more fossil fuels. Postwar nuclear energy capacity in the United States and the Soviet Union came with a special set of environmental concerns.

Mining is a spectacularly destructive activity, and the process of turning mined bauxite into aluminum is no exception. The efforts to develop and maintain operations to blast holes in the earth's crust and messily extract ore leave ruptured ecosystems, heavy metals, and sundry chemical contaminations, producing environmental damage on a scale that, in the words of historian Timothy J. LeCain, constitutes "mass destruction."[65]

Once the bauxite is mined, it needs to be refined into alumina by separating out other minerals from the ore. Sociologist Mimi Sheller estimated that four tons of bauxite produce two tons of alumina, which results in one ton of primary aluminum metal.[66] Sheller described bauxite mining as "an open pit process that leads to deforestation and leaves behind toxic 'red mud' lakes that can overflow and pollute local ground water. Bauxite mining damages forests, pollutes waterways, and encroaches on agricultural land often displacing small farmers."[67] The Container Recycling Institute's Jennifer Gitlitz identified the consequences of strip-mining bauxite as including soil erosion, water pollution, and habitat destruction. "Strip mining destroys whatever wildlife habitat has existed above the mine, and is difficult—if not impossible—to re-establish even with intentional revegetation."[68]

Ecologists and public health advocates have scrutinized bauxite mines throughout the world and found negative effects on human and animal health.[69] A representative study of locally produced food found high levels of heavy metals in Jamaican crops, sufficient to be "deleterious to human and animal health."[70]

In addition to heavy metal poisoning from ingesting crops grown near mines, the human health consequences of aluminum production include respiratory and circulatory diseases from inhaling alumina and bauxite dust, including cancers.[71] Broader environmental problems associated with the production of aluminum include about 1 percent of worldwide greenhouse gas emissions in general and the majority of emissions of the highly toxic greenhouse gases tetrafluoromethane and hexafluoroethane. Aluminum smelters release sulfur dioxide, fluoride, and spent pot lining into the air, land, and water.[72]

These problems existed in prewar aluminum production, but the vast expansion of capacity during the war magnified the environmental damage. The environmental historian Matthew Evenden identified the commodity chains of war-time aluminum as creators of significant damage not only due to the manufacture of aircraft in the United States and England, but also because of the razed forests and bauxite mines of Jamaica, the smelters and river dams of Quebec, and the shipment station of bauxite in Trinidad. He concluded that wartime aluminum production was "a revolution not simply of industry, but also of an expanded commodity trade, and of nature and society at an increasingly global scale."[73]

In British Guiana, the Demerara Bauxite Company's miners surface-mined bauxite, then washed it and put it on ships traveling through the Caribbean and up to North America. Smelters in Quebec processed this ore from the Caribbean, then shipped ingots for production to factories in the United States, Canada, Great Britain, and Australia.

This globalized trade's effects on ecosystems have included deforestation around the mines in British Guiana, destruction of fish populations disrupted by hydroelectric dams in North America, and the release of carcinogenic polynuclear aromatic hydrocarbons in waters around the dams, leading to raised cancer rates among aquatic mammals.[74]

North American operations during the war and in the decades afterward relied extensively on Jamaican bauxite. Since 1945, several studies of Jamaican mines have revealed significant threats to human and ecological health. In addition to respiratory illnesses, Gitlitz associated alumina spills with damage to coastal coral reefs.[75]

Soviet aluminum production traced similar international paths. Wartime bauxite mining in Hungary and Yugoslavia expanded after the war. By the end of the twentieth century, roughly one-quarter of Hungary's 3.4 million tons of hazardous waste was red mud from aluminum production.[76]

The Road to Sustainability?

In the twenty-first century, Alcoa, Reynolds, and Kaiser remain giants in aluminum production. In 2005, Alcan spun off an American producer called Novelis, now a subsidiary of India's Aditya Birla Group. Aluminum production is globalized, harnessing the energy of hydroelectric dams from Washington to India, resulting in further environmental degradation. The environmental and human health effects of bauxite mining and virgin aluminum production are being recognized as environmental justice issues and critiqued as neoliberal exploitation of vulnerable areas by global capitalism.[77]

Aluminum production is destructive to ecological and human health. That statement is not controversial. It does, however, raise a question about the building visited at the beginning of this chapter. How, knowing what we do about the ways aluminum is produced, could William McDonough incorporate aluminum in the construction of a sustainable building designed to behave like a tree? In order to address this question, it is necessary to understand how and where aluminum was used during the expansion of primary aluminum production between 1945 and the end of the twentieth century. Aluminum's applications shaped the visible postwar landscape in ways that residents of modern industrialized societies now take for granted.

Chapter Two

Designing Waste

Visitors to the artist Chris Jordan's website might be initially confused to see an image of Georges Seurat's 1884 painting *A Sunday Afternoon on the Island of La Grande Jatte*. The painting is a significant example of French neo-impressionism, but one normally finds reproductions of famous paintings in art history textbooks, on posters, or in catalog materials from their home gallery (in this case, the Art Institute of Chicago), not on a twenty-first-century artist's website. Why would Jordan display Seurat's work?

Closer inspection of Jordan's website reveals why *A Sunday Afternoon on the Island of La Grande Jatte* is there. The image is not actually a reproduction of Seurat's oil painting on canvas. The title is *Cans Seurat*, a work completed in 2007.[1] Clicking on the image zooms in to reveal that it consists of 106,000 aluminum cans organized to resemble Seurat's painting. That is the number of aluminum cans used in the United States every 30 seconds.

Jordan's depiction of Seurat's painting is part of a burgeoning field of environmentally themed art and fashion focused on issues of consumption and waste. Since the early 1990s, Ann Wizer's Virus Project has incorporated plastic waste from Indonesia into chairs, sculpture, and even clothing. In that spirit, "trashion" shows and exhibits featuring clothing created by ecologically minded designers like Timo Rissanen, Padmaja Krishnan, and Holly McQuillan from disposable materials have become frequent in New York City and other urban areas. I learned of the term in conjunction with a 2011 exhibit at Columbia College Chicago's

Averill and Bernard Leviton A+D Gallery, which displayed items such as a jacket made of discarded ski gloves, an obelisk made of 800 pounds of stacked discarded clothing, and a sculpture made of discarded dress shirts and a porch door.[2]

Much of this field and indeed much of Jordan's other work (including a 2011 version of *A Sunday Afternoon on the Island of La Grande Jatte* made of 400,000 plastic bottle caps) focus on the scale and consequences of the plastic garbage found all over the world. *Cans Seurat* is interesting because it reveals that aluminum became a disposable material in the years after its wartime mass production. How and why a metal that was so expensive to produce became waste relates to design choices capitalizing on a culture of convenience and disposal.

Fighting the Cold War with Aluminum

The two superpowers dominated the half century after World War II. The United States and the Soviet Union developed the largest military investments in technology in history, and aluminum played a large part in the form of new military aircraft and material for missiles. The scale and complexity of construction escalated as each side tried to make airplanes and missiles faster, more stealth, and deadlier. By 1948, the US Air Force had begun subsidizing machine tooling capacity to produce larger forgings of aluminum. Over the next decade, the air force invested hundreds of millions of dollars in coordinated efforts among the military, the Massachusetts Institute of Technology, aluminum producers, and military contractors to develop computerized milling operations in what Thomas P. Hughes identified as a military-industrial-university complex. Intercontinental ballistic missiles and airplanes from the massive B-52 to the stealth B-2 fed immense amounts of capital into military production. The year before the Berlin Wall fell, the US military paid $2.3 billion for each B-2 Spirit that Northrop Grumman produced (about $4.4 billion each in 2015 dollars).[3]

The Cold War was also fought on the home front, framed as a debate between communism controlled by a central government and free-market capitalism, which offered innovation and choice to what historian Lizabeth Cohen characterized as a nation of consumer-citizens. The debate overstated how free that free market was, since the federal government also acted as a consumer of military technology and large infrastructure projects. But the rhetoric grafted Cold War ideologies onto individuals' agency to consume goods made by corporations.[4]

This celebration of consumption was made explicit in July 1959, when Vice President Richard M. Nixon traveled to Moscow to debate the merits of the American economic system with Soviet premier Nikita Khrushchev. The exchange occurred during the second half of an East-West cultural exchange. One month before, an exhibition in New York City had displayed examples of Soviet technological achievements ranging from washers and dryers to space capsules and *Sputnik* satellites. In exchange, several American corporations sent examples of American technology for exhibit in Moscow. The American products ranged from automobiles to Pepsi, and the display included three fully automated kitchens with modern conveniences by General Electric and RCA Whirlpool and processed foods by General Mills. The American media sent several images of open refrigerators showing boxes, cans, and bottles of prepackaged food and beverages. As Nixon and Khrushchev sniped at each other's economy, technology, and even sodas, the American publicity materials for the exhibition declared that what was on display represented an average home available to all US citizens.[5]

Had it truly been from an average American home, the kitchen would have contained several examples of aluminum. Handles and the finishes on drawers and appliances might have been aluminum, and the presence of aluminum pots, pans, spatulas, and other utensils would have reflected the expansion of existing prewar industrial use. Most significant, if the kitchen were actually in use by an average American family, the garbage can would also have contained aluminum. Not the trash can itself, but its contents, since the packaging of many of the goods in the refrigerator was made of the metal. By 1959, what had been a scarce material was sufficiently abundant that it was subject to mass disposal.

The Age of Trash

The past was prologue. Aluminum had seen limited use in disposable products prior to World War II. The entry point into the aluminum market by Reynolds Metal was R. J. Reynolds's use of aluminum foil to line its cigarette packs and keep the tobacco fresh; somewhat more durable uses included affordable, reusable kitchen utensils. In the years between 1945 and 1970, aluminum became associated with trash and trashy things in a variety and volume far beyond that imagined before the war.

This growth was a matter of design. Aluminum producers worked with designers both in their employ and in other industries in the 1940s, 1950s, and 1960s to find new applications for their product. Many of those uses related to convenience, transforming what had been a coveted material into fodder for consumption and disposal.

Although aluminum had a reputation as a malleable and expensive material before the war, it gained a reputation as ersatz and inferior to other materials, as the

Aluminumware display, c. 1954. S. H. Kress & Company photograph album, Baker Library, Harvard Business School (olvwork691274)

historian Eric Schatzberg argued. "When aluminum competed with traditional nonferrous metals in industrial applications such as wiring, its chief advantage was typically price."[6] Despite aluminum being poorer at conducting electricity than copper, manufacturers worked to use the material in wiring from the late nineteenth century through to its expansion in more than 1.5 million American homes between 1965 and 1971. But aluminum wiring became associated with greater risk of fire. For Schatzberg, the history of aluminum wire reveals the change in how manufacturers and consumers saw the metal. "In the end, aluminum household wiring was marked symbolically as ersatz, a cheap—and in this case dangerous—substitute."[7]

"All the Traditional Beauty of Colonial Architecture"

Fire risks put popular attention on aluminum wiring as a household good (and problem) in the 1970s, but prior to that, the most conspicuous use of aluminum in the postwar housing industry was aluminum siding, rendered iconic in director Barry Levinson's 1987 film, *Tin Men*. The joke of *Tin Men*'s title is that the ethically challenged salesmen going door to door in 1963 working-class Baltimore, attempting to convince homeowners to class up their houses with their wares, weren't even selling tin, but a representation of tin actually made of aluminum.

Metal siding was a new development in the postwar economy. A pioneer in the process was Hammond, Indiana, machinist Frank Hoess, who in 1939 secured a patent for an experimental steel siding designed to resemble wooden clapboard. In 1946, Hoess partnered with a new company, Detroit's Metal Building Products. Metal Building Products was an initiative organized to promote and sell Hoess's siding, but made of Alcoa aluminum rather than steel. By the end of that year, the siding was installed on several new housing developments in the northeastern United States.[8]

Although Metal Building Products failed after two years, the proliferation of aluminum siding had just begun. The growing postwar housing industry adopted aluminum siding, which was promoted as a modern facade that served to effectively insulate homes. Reynolds Metals, as part of the company's attempts to expand the market for aluminum, initially attempted to mass-produce houses with aluminum frames and walls. A 1946 plan to build a subdivision of aluminum houses in Louisville, Kentucky, was stymied by a failure to secure building permits from the city, and the company then focused on marketing the components of its aluminum houses instead of attempting to create developments. By the end of the summer of 1946, Reynolds had begun to market aluminum siding and roofing shingles.[9]

Unlike Metal Building Products, Reynolds was able to market its products nationwide. In 1947, it purchased full-color advertisements in the *Saturday Evening Post* touting aluminum siding as providing "all the traditional beauty of colonial architecture" with modern insulation, fire protection, and the added incentive of easy care by homeowners. "Think of year after year with no maintenance, for aluminum needs no painting." By the time Reynolds ran the ads, it boasted of selling enough siding for 141,113 homes nationwide.[10]

Reynolds was not the only company to experience success with aluminum siding. The Kaiser Aluminum and Chemical Corporation also desired to expand the postwar market for the metal. During the war, Kaiser had quickly built temporary housing for its wartime workforce, and the combination of that experience and federal subsidies to address the acute postwar housing shortage produced designs for prefabricated housing employing aluminum. Two days after Japan surrendered to end World War II, Kaiser partnered with a Los Angeles real

550 South 20th Street (house), Louisville, Jefferson County, KY, 1982. Aluminum siding, screen doors, and fences became common features of American vernacular architecture in the postwar era. Photograph by D. Mitchell, Historic American Buildings Survey, Library of Congress (HABS KY-190)

estate developer, Fritz Burns, to build at least 10,000 such homes on the West Coast.[11]

As Kaiser planned its prefabricated housing, it found Canadian inventor Charles Kinghorn, who proposed a curved sheet-metal clapboard. Kinghorn applied for a patent on his design in January 1947, and shortly thereafter Kaiser purchased exclusive rights to manufacture and market his clapboard. In April 1948, Kaiser's Trentwood, Washington, plant began producing Kinghorn's siding. Kaiser developed a national network of contractors to use the new siding and took advantage of its relationship with large-scale developers. At a time when a contract with a developer could lead to materials being used on thousands of individual buildings, such relationships could determine the success of a product. The Kaiser siding became a common facet of new subdivisions on the West Coast.[12]

Also in 1947, Akron, Ohio, developer Jerome Kaufman got into the aluminum-siding business, founding Alside Incorporated that spring and developing a dealer network throughout the Midwest. Alside's approach was novel in that it sold pre-painted aluminum siding, allowing homeowners the advantages of aluminum insulation and the look (more or less) of traditional white clapboard homes. This aesthetic innovation proved popular; by the end of 1948, Alside had recorded gross sales of more than $1 million (about $10 million in 2015 dollars). Its competitors Reynolds and Kaiser joined in offering prepainted siding by 1949. As the 1950s began, aluminum siding had become a mainstream aspect of the American construction industry, offering a simulacrum of wood to middle-class homeowners. Its success was instructive to aluminum producers as they worked to expand their market.[13]

Tin Men is a nostalgic portrait of aluminum siding as ersatz. The aluminum skin of Pittsburgh's Alcoa Building had garnered praise for showing the gleaming metal's aesthetic appeal; in the words of historian Stuart W. Leslie, it was an example of "architecture parlante," a building that speaks of its function and meaning. (Some critics of the time did not see the appeal. In 1954, the *Baltimore Sun* concluded that the Alcoa Building "looks kind of funny, like a Florentine palace in the wrong shape and wrong materials.")[14] Conversely, painted or vinyl-coated siding did not intend to show the metal's properties. Instead it was meant as a durable, affordable substitute that resembled wood. Unlike aluminum electrical wiring, aluminum siding was not associated with hazards beyond injuring aesthetic sensibilities.

Designed for Disposal

Aluminum siding was reasonably durable, lasting years or even decades before renovations rendered it waste. Its mass production came at a time when the design departments at all three major American aluminum producers were working to expand applications of the metal to goods that had limited durability, including products designed to be disposed of after one use. Reynolds's Styling and Design Department, Alcoa's Market Development Department, and Kaiser Aluminum's Industrial Design Department all worked with companies that were potential buyers of aluminum.[15] If more industries understood, in the words of design historian Dennis P. Doordan, "the properties and performances of aluminum and aluminum alloys," the market for the metal would expand. Thus the design departments of the aluminum producers worked "to encourage other designers to find new uses for aluminum."[16] Alcoa's Frank McGee declared that the company's design division aimed to put its "knowledge and facilities at the disposal of any designer who can use them. Our own industrial design group was established for the sole purpose of aiding industrial designers in their projects—not in any way to compete with the designer!"[17]

These design divisions openly promoted aluminum use in meetings, lectures, and publications. Reynolds published two volumes of *Aluminum in Modern Architecture* to expand the market for buildings like Alcoa's aluminum-clad headquarters—as well as a variety of products to fill those buildings. "The fullest exploitation of anything new whether it is an idea, a technique, or a material," John Peter wrote, "demands understanding and imagination."[18]

Alcoa also produced a pair of texts intended to expand aluminum use. *Design Forecast 1* (1959) and *Design Forecast 2* (1960) reveal how extensively aluminum had already made its way into a variety of uses since the war. The publications existed, McGee declared, because as Alcoa "became aware of the dynamic importance of the industrial designer, our conclusion was inescapable. Since we are producers of a material that must be chosen and specified, we must be in contact with those who are doing the choosing and specifying. Possessing a wealth of information about our metal, we must see that it is available to those who can use this knowledge."[19]

One aspect of that assistance was teaching designers that aluminum was not simply one metal with uniform properties but, like steel, a variety of alloys that had relative merits and limitations. *Design Forecast 1* provided a chart of various alloys as a quick guide to their existing uses, their properties, and their potential. Alloy 1100 was described as having "excellent resistance to corrosion" and was often found in spun hollow ware and decorative parts. Alloys 2011 and 2017, found in screw machine parts, had only fair resistance to corrosion but were described as having excellent strength and machinability.[20]

Alloy 5052 was described as having a "good combination of strength, corrosion resistance, [and] finish" and was often used in appliances and transportation uses. Similarly, 5457 was often used in appliances and auto trim because of its "excellent finish characteristics." Alloy 6463 shared 5457's finish characteristics and was also used in auto trim.[21]

Alloy 5056 was very strong, resisted corrosion well, but had poor machinability. It was often used in screen cloth, wire products, and fencing. Another alloy used in fencing was 6061. Its "good strength and weldability" made it suitable for furniture as well.[22]

Due to its "high corrosion resistance, good appearance, and low cost," 6063 was often used in windows and storefronts. Alloy 7075, the origins of which were discussed in chapter 1, was described as having "very high strength and hardness." In addition to its use in aircraft, it was also often found in keys.[23]

Alcoa emphasized that although some alloys were rated stronger or easier to work with than other alloys, all of the ratings it offered were relative to other aluminum alloys and not to other metals. Alcoa then stated that all aluminum alloys were among the most corrosion resistant of metals, and even the alloys rated lowest on durability were "frequently used unprotected with totally satisfactory results."[24]

In *Design Forecast 2*, *Industrial Design* editor Ralph Caplan led a roundtable of industry professionals in a discussion of the metal's uses and relevance to the American economy. Caplan argued that a case could be made for expanding aluminum's use in disposable products. "We have an economy in which it makes some sense to discard things," Caplan said. "Carrying this further, if we had an insular situation, were completely separate from the rest of the world, it might even become moral to do so; perhaps immoral *not* to discard things." He then pulled back slightly from this stance, noting: "We are developing a physical closeness to the rest of the world, and it does not have these standards."[25] Caplan argued that by 1960, a stewardship of owned goods, "a certain kind of desirable materialism," had disappeared in favor of a culture of disposability. "We are losing pride in specific ownership: the car whose fender we used to pat, the pocket knife that our children used to fondle so carefully and proudly."[26]

In response, Earl F. Bennett, the manager of architectural sales for the Koppers company, declared: "There are five simple statements that describe the traditional approach to a sense of values of our country; use it up; wear it out; make it do; go without; we choose how we spend our money. These have been with us a long time, and still are important to us."[27]

Charles E. Whitney of Whitney Publications argued that designers had made the modern world better and easier to live in. "Our great changes in the convenience and ease of living of the past four decades have given man a whole new set of values and interests. The designers and the industries they serve have made this possible. If you were one of those who observed the life of our parents forty or more years ago, you would never exchange, nor would you condemn, the 'things' that have altered our lives."[28]

One result of these changes was an array of goods created for the convenience of postwar consumers. Convenience meant that goods were affordable, and in some cases affordable imitations of more expensive goods. Convenience also meant that when a product had been sufficiently used by the consumer, it could be disposed of without consequence to the consumer.

That same year, Vance Packard decried this state of affairs in *The Waste Makers*. An economy based on creating cheap, disposable goods designed to break down, Packard argued, risked "the dangerous decline in the United States of its supply of essential resources."[29] Aluminum, that expensive and coveted prewar material, was now the stuff of trash. "Steaks and other meats have appeared in disposable aluminum frying pans," Packard noted. "When the steak is done, just throw away the pan along with the nasty old grease."[30]

Packard laid responsibility for this culture of disposability at Alcoa's feet. "A sales executive at the Aluminum Company of America announced that the day was at hand when packages would replace pots and pans" in the preparation of meals.[31] Indeed, the proliferation of TV dinners during the 1960s was enabled by the use of aluminum to make packaging that could be placed in the oven, heated, served, and then disposed of when the meal was completed.

Packard said that mass-produced goods made of aluminum had declined in quality over the 1950s. Lawn furniture made of hollow tubes was, he argued, experiencing "a sharp downtrend" in quality, producing "many angry outbursts and defenses in the trade press during the late 1950s. One store owner complained: 'Standards have gone to the winds.'"[32] Packard referenced a litany of dealer complaints, including claims that manufacturers were using thinner gauges of aluminum than they had earlier in the decade; that the mesh webbing that formed the seats of the chairs was thinner and weaker so that it "would quickly give way when people weighing more than 140 pounds sat on it"; and that furniture that had once been manufactured with stainless-steel bolts now was joined with aluminum rivets.[33]

Not that these chairs were ever designed for durability; they were created to be mass-produced at low cost.[34] The historian of technology Phil Patton traced the origin of the aluminum lawn chair to the aluminum tubing used in military aircraft frames, noting that Alcoa called the process of finding new product applications for processes developed during the war "imagineering."[35]

Critic Craig Vogel called the first aluminum folding chairs "an instant success" because they "were easy to maintain and light enough that any member of the family could move them." Vogel cited a 1954 advertisement for the Totalum chair, which targeted female consumers, declaring: "There's no longer any need for the male member of the family to flex his muscles when it comes time to set up the folding chairs for lawn parties, bridge parties or outdoor meals."[36] The portability of the chairs enhanced their convenience. They could fold flat and "be taken to the town picnic, the lake, the beachfront, or to Little League games."[37] Ease of manufacture, portability, and low price meant these chairs quickly became mass-market goods sold in great quantities at drugstores and department stores.[38]

Vogel noted that these chairs were among the most popular ever sold, but not among the "best chairs ever designed." The aluminum folding chair's economic efficiency

> far outweighed its appearance (aesthetics) and level of comfort (human factors/ergonomics). While its low cost, light weight, and portability made it a big seller, it had several serious design flaws that made it uncomfortable, unstable, and difficult to repair. The material used for the seat and back was often striped in combinations of green, white, orange, blue, and red plastic. Weaving the stripes produced a garish plaid that visually overwhelmed the minimal aluminum tubular frame, which itself was easily damaged. While its light weight was an advantage for mobility, it was not ideal for stability—the chair easily tipped to one side, and even a small gust of wind could blow it around the backyard. The connectors holding the chair together were the cheapest, off-the-shelf-solution, and there was little attempt to match the details to the whole. In sum, the aluminum folding chair was a group of parts assembled to respond quickly and inexpensively to an opportunity in the marketplace.[39]

Packard's complaints about declining quality assumed that consumers cared about the design or quality of aluminum folding chairs in the first place. Discussing these chairs, Vogel cited Donald Norman, the author of *The Design of Everyday Things* (originally published in 1988 as *The Psychology of Everyday Things* and subsequently retitled for editions published in 2002 and 2013), in arguing that humans have learned to adapt to poor design. "They often see themselves, rather than the object, as the problem. This is certainly the case with the tubular aluminum folding chair."[40]

Food Packaging

Aluminum folding chairs might be uncomfortable, cheap, and frail, but they are designed to function repeatedly for the consumer. Other applications are designed to be used only once before disposal. Single-use aluminum food and beverage packaging had by 1960 become symbols of what Vance Packard called "the throwaway society" in the United States. In 1961, John A. Kouwenhoven captured the lack of concern about waste by titling his collection of essays on "what's American about America," *The Beer Can by the Highway*.[41] Less amused was the German-born landscape architect Peter Blake, who remarked in 1964 that commercial waste was despoiling nature and injuring the nation. "In destroying our landscape, we are destroying the future of civilization in America."[42]

Design for disposability aroused Blake's ire and Kouwenhoven's amusement. Packard shared Blake's angst over cheap, disposable aluminum products; aluminum manufacturers celebrated the business opportunity. Having established a market for cheap aluminum furniture, they saw the next attractive set of convenience products to be packaging for food and beverages. In *Design Forecast 2*, Alcoa championed the use of its product in single-use disposable food containers. "The post–World War II appearance of the foil baking pan was the initial use of the rigid foil container. On its heel have come TV dinner and other heat-serve packages, [and] decorative containers for items as diverse as food and flowers."[43]

Alcoa's competitors, especially Reynolds, had made inroads into this market with aluminum foil, which consumers shaped to fit leftover foods, and prefabricated containers for TV dinners and other food products. Reynolds Wrap became a brand name as omnipresent as Kleenex as aluminum foil became a common presence in postwar kitchens. Design historian Craig Vogel noted that aluminum became a staple of affordable kitchenware in the 1950s. "In 1950, Heller Hostessware (1946–c. 1955) introduced a set of aluminum dishware called Colorama. The tumblers for this line were deep drawn and anodized in a rainbow of colors."[44]

In the 1950s and 1960s Alcan, Alcoa's Chicago Metallic Division, and the E-Z Por Corporation of Niles, Illinois, marketed "reusable, disposable" aluminum trays "stamped with intricate designs that closely simulate[d] more expensive serving dishes." E-Z Por also filed a 1967 patent on an aluminum broiler pan, "which may be inexpensively made and which may be discarded after each use and does not have to be cleaned. There is therefore provided a throw-away utensil which is very inexpensive and economical."[45]

But E-Z Por's application was for a product with existing competition. By 1961, Kaiser had filed a patent on an aluminum tray "for meats or similar article[s]," and Arnold G. Keppler of Seattle had filed a patent for a broiling tray, "which is made of such lightweight aluminum that it can be used and disposed of along with all the grease that has been rendered from the foods being broiled."[46]

Almost as an aside in 1960's *Design Forecast 2*, Alcoa declared that "the long-heralded aluminum can became a fact during the last two years."[47] Vogel credited the Colorama dishware with providing the technical expertise to create aluminum cans. "While the Colorama series proved a less than optimal solution for a housewares product, it set the stage for one of the most effective uses of aluminum in the twentieth century: the aluminum can. Aluminum manufacturers perfected the process of deep-drawing aluminum to form containers."[48]

Cans made of steel and tin had been important to industrial food production since the French engineer Nicolas Appert developed a method for hermetically sealing food in glass jars in 1809. Sealed metal cans allowed for the preservation and transportation of foods and liquids, including military rations. Steel was especially useful due to its strength, which allowed the American Can Company to make a flat-top beer can in 1935 without fear that the pressurized liquid would damage the container.[49]

The process of making aluminum cans ubiquitous on the American landscape took time. Two hurdles existed. One was the use of steel as a material to create beverage cans. The second was the long-standing use of glass bottles. Glass, reflecting its history in Appert's innovation, had been used for beverages since the nineteenth century. How it was used and how it eventually was surpassed as a beverage container reflects the changes in the beverage industry and consumer tastes; producers gradually shifted the burden of packaging away from themselves and toward the people who purchased their products.

Beverage companies that used heavy glass bottles to distribute their products in the early twentieth century considered the containers too expensive to give away with the beverages. The historian Robert Friedel wrote that bottles from this period often came with notices on them, such as "This bottle to be washed and returned" or "This bottle not to be sold." A widespread example of this conception of the bottle was in sales of milk. Dairies used glass bottles to distribute milk. Since many dairies were small operations, they used a limited number of milk bottles, which were returnable. Refilled containers made a circuit between the producer and the consumer, interrupted only if the bottle was broken.[50]

Glass had several benefits for bottlers. It did not alter the taste of the beverage, and transparent glass allowed consumers to see the quality of the product. Milk had come under suspicion due to widespread scandals in the nineteenth century concerning the adulteration of milk with chalk and even extracting milk from deceased cows, a product known as "swill milk." Such schemes led to federal regulation (including the passage of the 1906 Pure Food and Drug Act); glass bottles reflected manufacturers' attempts to instill consumer confidence in their product.[51]

This relationship began to change around World War I, in part because breakage was a problem plaguing glass bottles. Instead of replacing the returnable glass bottle, however, beverage companies transformed it into a smaller, cheaper vessel that could be lost with negligible cost to the bottler. Coca-Cola became a dominant producer of soft drinks in part due to its iconic glass bottle.[52] Unlike dairy bottles, the Coke bottle was designed as single-use, disposable packaging material. It was a smashing success. Coca-Cola's 6.5-ounce disposable bottle dominated the soft drink market for 40 years, gaining significant competition only in 1955, when Pepsi produced a larger 10-ounce disposable bottle.[53]

Glass bottles also were used for beer, though by the 1950s steel competed as a suitable packaging material. The advantages of glass were that consumers could see the product, and it did not produce an aftertaste. The disadvantages of glass included its excessive weight and fragility. Transportation and breakage contributed to the operating costs for beverage distributors. Steel, conversely, was lighter and more durable, allowing beer, beans, and other canned edibles to be stored for long periods of time. Packaging historian Thomas Hine noted, "Putting soft drinks in cans was an obvious solution. Canned beer had been found to occupy 64 percent less warehouse space than the same quantity of bottled beer and the shipping weight was less than half as much."[54]

Being lighter than steel, aluminum was an attractive alternative as a canning material. The process of replacing steel, however, took several years. Kaiser Aluminum introduced the first aluminum beverage can in 1956. Its first major use came two years later, when the Coors Brewing Company introduced its seven-ounce aluminum cans. Coors spent $10 million over five years to develop a can appropriate for beverage distribution.[55]

Once Coors's innovation was successful, other companies followed suit. RC Cola began using 12-ounce aluminum cans in 1965, and two years later Coca-Cola and Pepsi adopted aluminum cans, offering more durable alternatives to their iconic glass bottles. A decade after Kaiser had introduced the aluminum can, it had gained a foothold in the American beverage industry.[56]

Steel was stronger than glass, but it had disadvantages of its own. Steel was heavy, and consumers found opening the cans to be cumbersome. Aluminum initially was used to reduce the weight of steel cans and make them easier to open. The point where aluminum was used was the sealed opening at the top of the can, sometimes in conjunction with an all-aluminum can top fused to the steel that made up the rest of the can. Ermal Fraze's pull tab, for which he received a patent in 1963, eliminated the need for a can opener. Vogel credited Fraze with a major engineering accomplishment. "Engineers had to create a sealed container that could be opened by hand with a very low incidence of failure. The fact that pure aluminum cannot be welded posed a major challenge. The solution—a pull tab attached to a pin on the lid—placed incredible demands on manufacturing tolerances. The scoring of the lid had to be thin enough to pull away, but strong enough that it wouldn't explode under pressure."[57]

Vogel identified problems with this initial use of aluminum in cans, some relating to functionality and others contributing to the waste Packard bemoaned. "The first pull tab had a few flaws, resulting in an unacceptable failure rate. The handle could break away completely or remove only part of the aluminum top. Once opened, the pull tab, which separated completely from the can, was discarded, and it quickly became a nuisance. Its sharp edge was dangerous and it was littered everywhere. In 1973, the pull tab was replaced by a push-through tab, which remained connected to the can."[58]

Although consumers initially preferred glass, in part because aluminum alters the taste of beverages, consumer preferences changed as the experience of purchasing food and beverages evolved in the postwar economy. By the end of the 1950s, food shopping patterns had changed dramatically. Shopping became associated with automobile culture as suburban subdivisions proliferated. Urban deliveries from neighborhood dairies and grocers gave way to customers driving their new cars to large supermarkets, where they bought large quantities of food and then drove home.[59]

Consumers began to buy beer and soft drinks by the case. Glass bottles had to be returned and reused, a major responsibility for soft-drink companies and an annoyance for customers. The reuse of glass bottles was finite, required careful sterilization, and was complicated by the material's tendency to break. Bottles were heavy, delicate, and inefficient to transport.[60]

The year 1963 was an eventful one for aluminum in cans. Fraze received the patent on his aluminum pull tab, and a new method of producing all-aluminum cans allowed the metal to better compete with steel. The drawn and ironed method of production involved punching a cup out of a metal sheet, at which point the fabricator drew out and ironed the sides of the cup until they were about one-third of their original thickness. The result was a strong, light aluminum can that was cheap to make.[61]

Vogel explained aluminum's expanded use in beverage containers as a material advantage. "In contrast to the glass bottle, the aluminum can is a highly efficient container. The weight of the can does not add significantly to the total product weight when filled," and aside from saving on transportation costs, aluminum cans facilitated transportation and storage because of their suitability for stacking.[62] A 24-can case of soda or beer takes up far less space in a truck or store than the equivalent number of glass bottles does, and six-packs in each material reveal that aluminum cans require less space. As beer and soft drink companies expanded their businesses, aluminum cans allowed them to scale up production and distribution beyond what glass bottles could.

Inventing the American Litterbug

As single-use packaging reshaped the American waste stream, a problem for the food and beverage industry was that this created conspicuous trash branded with corporate logos. Much thought went into the design of wrappers, trays, cans, and pull tabs to reduce transportation weight, preserve contents, and look appealing. No thought went into the fate of the packaging once consumers had exhausted the contents. The waste created by designing for disposability was so recognizable in industrial societies that a 1970 article on what math teachers could do to teach ecology included several sample problems about automobile pollution and the question: "If 48 billion aluminum cans are used annually in the United States and there are 207 million people, then how many cans are used per person per year?"[63] The fact that math teachers could use "billions" when referring to these cans in 1970 without disbelief is evidence of the scale of the problem. When consumers threw a Coors can or a McDonald's wrapper on the ground, observers could place the blame for the unwanted debris at the companies' feet. Which is where the responsibility originated.

Food and beverage manufacturers responded to the problem of conspicuous waste not by designing the trash away, but by engaging in public relations to shift any burdens relating to waste away from themselves. In the years before *The Beer Can by the Highway*, *God's Own Junkyard*, and *The Waste Makers* were published, American beverage and packaging companies moved to ensure that the responsibility for the effects of their packaging would shift to other parties. They established a trade organization called Keep America Beautiful (KAB) in 1953. KAB defined the problem of single-use, disposable packaging as litter, the result of aberrant consumer behavior rather than industrial design decisions. A historian of Coca-Cola, Bartow J. Elmore, described KAB's greatest strength as presenting itself as an independent third-party organization "interested in public service, rather than a corporate lobbying agency with a specific agenda to protect big business."[64] Describing KAB in 1954, the *New York Times* reported that corporate "intelligent self-interest" had prompted the new organization, which worked to persuade Americans that individual consumers, rather than corporations, should be responsible for disposing of corporate-created waste.[65]

KAB campaigns are evident today on television, on billboards, and in print advertising; these are continuations of the broad public relations campaign KAB began in the 1950s, which included hundreds of print and television advertisements training consumers to bear the burdens of waste. Elmore noted a 1960s television spot using images of otherwise scenic settings for picnics, swimming, and camping that were besmirched with trash. Showing the befouled scenes, KAB explained, "One thing's sure, America's litter problem is in your hands," adding, "Keeping America clean and beautiful is your job."[66]

Postwar consumption and waste created new cultural phenomena. The nineteenth-century street was, in addition to being a conduit for transportation, also a designated sink to dispose of food waste and other household discards. As streets transitioned from dirt to wood, then to cobblestone and asphalt, they were increasingly perceived as exclusively for transportation. By the late 1940s, people tossing waste in the streets were branded "litterbugs," and KAB popularized that insult, which equated individuals throwing trash on the ground with vermin. Elmore noted that "litterbugs, as KAB's promotional pitch went, were in many ways subhuman, akin to disease-ridden insects that were the target of pesticide campaigns in postwar America. They had to be eradicated."[67]

By the end of the 1960s, America's embrace of disposable packaging rendered returnable glass bottles into historical artifacts while aluminum foil, disposable trays, and empty cans filled garbage cans and landfills. KAB's advertising pushed the notion that consumers seeking convenience were responsible for the consequences of waste; privately, executives at food and drink companies admitted their culpability. Coca-Cola president Paul Austin acknowledged this fact in 1968. "We participate in the [creation of] litter to a significant degree," adding that the company had "earned various criticisms for littering the landscape." Aluminum's durability represented an image problem for Austin's employer; he lamented the fact that "the packaging for our products is highly visible" and expressed frustration that Coke's "colored decoration on a can or the unique shape of our bottle doesn't deteriorate as readily as paper containers."[68]

Disposable aluminum cans, however, were simply too profitable to give up. Coca-Cola and Coors found them useful for distributing their products, and aluminum manufacturers found them a terrific, constant application for theirs. Austin noted in 1970 that there were not "any serious prospects for truly degradable soft drink containers. Not now, anyway."[69] Instead of providing consumers with packaging that would not create litter, the companies' disposable food and beverage containers proliferated. Instead of designing to address the problem of waste, the trash's creators focused on public relations campaigns to shift culpability.

Their work was remarkably successful. By 1970, American consumers had been trained to purchase disposable containers and then place them in the garbage. Children watching Saturday morning cartoons were warned away from the perils of litterbugs as often as they were warned by Smokey the Bear to not start forest fires. Litter was now a product of individual behavior rather than the result of conscious design choices by large corporations. Having shifted consumers' behavior, food and beverage producers reshaped the waste stream. One early 1970s survey found that 5 percent of the solid waste in the United States consisted of aluminum cans.[70]

Having succeeded in their public relations strategy, companies elaborated on disposable aluminum food packaging beyond foil, TV dinners, and soda cans. In the late 1970s, the Ludlow Corporation and the American Can Company filed separate patents on disposable, single-use condiment packets fashioned from aluminum and plastic.[71] Heinz, Hunt's, Hellmann's, and other makers of ketchup, mustard, mayonnaise, relish, and other condiments adopted the hybrid-material packets, distributing them to fast-food restaurants, cafeterias, and sports stadia. The tiny packets added to the volume of aluminum in the trash; they were so lightweight that wind easily blew them out of trash cans and onto the ground. More than 30 years after affordable aluminum had been made available for commercial uses, designers had found more ways of turning the once scarce metal into garbage.

The Waste Makers

In the years after World War II, aluminum producers actively sought to broaden the market for their material by putting it in single-use, disposable products. Despite the ecological damage wrought by mining bauxite and producing primary aluminum, the massive energy infrastructure and commodity chains developed to create the metal during World War II meant that producers had sufficient capacity to continue making aluminum for these single-use purposes in the postwar era. Aluminum became a growing sector of the industrialized world's solid waste stream and increasingly visible as litter polluting roadsides, parks, and other areas.

The consequences of this waste varied with their design. Pull tabs from cans posed vexing dangers when left in the wild, where animals could be cut or could choke on them. By the time beverage companies moved away from pull tabs amid criticism, discarded condiment packets posed new choking hazards to wildlife. Small, light pieces of metal made their way into the environment through careless disposal by consumers or by being jostled free from garbage cans or during transport to landfills.[72]

KAB's shift of the burdens of this waste away from its corporate creators to individuals and municipalities came with an added cultural dimension. The "litterbug" creates trash not due to a biological drive, but due to poor morals and manners. KAB's demonization of consumers who littered brought a strong moral dimension to the issue of waste. Environmentalists by the 1970s bemoaned aluminum waste in ethical terms, and the continued use of an assault on morals is reflected in the recent spate of trashion and environmental art, such as Jordan's 2007 depiction of Seurat's painting.

Jordan's art is also evidence of attempts to reuse aluminum waste, attempts that have taken both artistic and industrial turns that critique aluminum discards even as they inform a new championing of reclaimed aluminum as an environmentally responsible material. The green reputation of aluminum is directly at odds with the stories of the metal in the first two chapters of this book; understanding the change requires discussion of the historical development of recycling in industrial societies.

Chapter Three

A Recyclable Resource

Brooklyn's Sunset Park neighborhood is home to the world's most beautiful recycling center, the Material Recovery Facility (MRF). Designed by architect Annabelle Selldorf's firm (which also designs art galleries, museums, and libraries in Europe and the United States) and built at a cost of $110 million, the MRF is a joint venture between New York City, which spent $60 million of local taxpayer money on the facility, and Sims Metals, one of the world's largest scrap metal dealers, which spent $50 million on the site. The L-shaped warehouse facility includes, in the words of the *New York Times* architecture critic Michael Kimmelman, "a public-friendly zone, with [an] education and visitors center, trees, bioswales, a grassy entrance and parking for school buses. The center, including offices, a cafeteria, classrooms and a terrace with a killer view over the harbor, became a light-filled, three-story shoe box, parallel with the pier."[1]

Eight miles northeast of this architectural marvel, Brooklyn's Greenpoint neighborhood is home, at least until gentrification replaces them with condominiums, to several small scrapyards. None are architecturally significant; none feature grass, classrooms, or "public-friendly zones"; and they work at volumes dwarfed by Sims's operations. What they share with the Sunset Park facility, however, are daily contributions to the New York–Newark area's significance in global scrap recycling. A pedestrian walking along Provost Street cannot help but experience the noise, dust, and traffic of these businesses chopping, baling, and sorting aluminum cans, wire, and construction materials discarded by consumers or businesses. Many scrapyards dot the periphery of New York City's Newtown Creek Wastewater Treatment Plant, making the area a cluster of waste management of everything from sewage to scrapped automobiles to beer cans.

The sources for American secondary aluminum production are a mix of public collection systems and private enterprises. In New York City, the Department of Sanitation collects curbside recycling from the city's neighborhoods and then trucks the material to the MRF run by Sims in Brooklyn, where it is sorted and baled. Sims then transports the recovered aluminum from the MRF to its facilities across the water in Jersey City. From there, secondary aluminum is sold to producers in the United States and other countries, going across North America in trucks and across the Pacific Ocean on container ships.

Sims and the smaller Greenpoint businesses are products of a history of salvage as old as the industrial revolution. Businesses trading in old metals and textiles grew as industrial production grew, connecting scrapyards in Brooklyn to brokerages in New Jersey, and fostering secondary material trades linking North America to Europe and Asia. Every day, scrap metal collected in New York City goes through a chain of trades via truck, train, and boat. Some metal finds its way to smaller mills in Texas and Mexico. Other metals are sold to China. Much like the bauxite ore business, scrap recycling is a global concern.

Since the paper-making business stopped using rags as a primary raw material in favor of wood pulp in the late nineteenth century, iron and steel have been the most widely traded secondary commodities in the world. Considering how much of the modern world is made of iron and steel, including buildings, transportation systems, and other infrastructure, the ubiquity of ferrous scrap should not be surprising. By the end of World War I, the journalist George H. Manlove noted that the global trade had exceeded $1 billion annually (more than $15 billion in 2015 dollars).[2]

Aluminum lacked the ubiquity of iron and steel in the early twentieth century. The story of its reuse is more recent. The mass production of virgin aluminum during World War II created an abundance of the metal in the United States and Europe. As discussed in the previous chapter, between 1940 and 1970, aluminum use spread from aviation to beverage containers and siding for houses. The metal gained a reputation as ersatz, cheap, and disposable, despite the environmental toll inherent in its creation.

The name "aluminum" represents a wide range of alloys. During the melting process, different alloys may be fashioned by adding elements such as copper, zinc, magnesium, silicon, and manganese either individually or in specified combinations. Those combinations produce different alloys, with the most common standards being 6061 (used in beverage cans and aviation), 7075 (used in aviation), 1100 (used in cooking utensils), 6063 (used in furniture and architectural details), and 2024 (used in aviation). Production and consumption begat disposal. Many of the products created with aluminum had limited functional lives. Lawn furniture broke. Oil cans were emptied. Screens tore. Siding cracked. Soft drinks were consumed.

What happened to these obsolete aluminum goods once their useful lives ended? One answer is they joined food wastes and other packaging materials in landfills. A second answer is they were recycled.

What it means to *recycle* requires elaboration. Much of modern society's historical understanding of recycling is that it was a movement that emerged in the wake of environmental concerns about trash at the end of the 1960s. This history of recycling focuses on drop-off centers, curbside collection, reverse vending machines, and the actions of consumers, who sorted materials in an effort to be environmentally responsible.[3]

This is an important history. But it is only a partial account of recycling. For these collection efforts to work, some demand for the collected materials had to exist. The longer history of recycling involves the scrap and salvage industries that evolved alongside heavy industry during the nineteenth century. Rags and metals were transformed from wastes to commodities as paper manufacturers, railroads, steel mills, and other industries found the salvaged materials to be affordable alternatives to extracting primary commodities. A Pittsburgh steel mill operating in 1910 would have found purchasing a ton of scrap iron from a local dealer cheaper than operating an iron ore mine hundreds of miles away, then transporting the ore back to its mill. The history of recycling is in some aspects the history of the industrial revolution.

Aluminum joins this narrative, with its own periodization. Unlike iron, steel, copper, and rags, aluminum was not a commodity of mass consumption between the mid-nineteenth century and the end of World War I. Scrap and salvage companies had been part of the American economy since colonial times, with cotton and linen rags initially the most widely traded materials. At the end of the nineteenth century, ferrous metals became the most widely traded secondary materials around the world, with the American trade alone worth more than a billion dollars per year during World War I. Demand from the American, British, German, and Japanese militaries for ferrous scrap spurred the growth of scrap iron and steel brokerages during World War II, and the largest salvage firms in the postwar era enjoyed steady contracts with the largest manufacturing companies, such as Ford and US Steel.

Aluminum was not a major sector in the secondary materials market during World War II. The low scale of aluminum production prior to the war meant that little secondary material formed the basis of production in Europe, Asia, or North America. However, with the expansion of global aluminum production, the supply of secondary aluminum in the form of both prompt scrap (excess metal produced in smelters and aircraft manufacturing) and market scrap (from junked aircraft) was substantial.

The mere presence of secondary material was insufficient for scrap aluminum to become a viable commodity, however. Salvage dealers needed to develop methods to ensure that the scrap sold to manufacturers was material the manufacturers could use. Metal dealers since the 1920s had processed copper, iron, and steel by cutting the ferrous metal away from other materials with shears and torches, using magnets to quickly harvest the desired materials, and then melting them in furnaces to burn away impurities and create new alloys. As primary aluminum production increased and as aluminum goods from airplanes to screen doors were designed to become functionally obsolete in the postwar era, aluminum recycling joined the mainstream of industrial commodities. By 1950, scrap comprised about one-third of all aluminum used in production in the United States; 10 years later, scrap comprised more than half of domestic production; and in the twenty-first century the proportion of scrap in aluminum production ranges between 55 percent and 60 percent.[4]

The story of scrap aluminum shares much with the stories of earlier scrap commodities, but it also has distinctive aspects of materiality and timing. World War II marked the largest application of resources to military functions in the history of the planet. Metals of all kinds were employed, as were energy sources of all kinds, which were joined by textiles, rubber, and new synthetic uses of petroleum. As military equipment was damaged or worn, scrap dealers throughout the world salvaged battlefield losses and obsolete inventory in volumes surpassing the peacetime collection of materials or even the military bounty of World War I (when annual ferrous scrap sales surpassed a billion dollars worldwide for the first time).

Military Scrap

Scrap aluminum fed the market that expanded due to government and industry collaboration. However, the direct intervention of the state in the large technological systems producing aluminum after 1930 was considerably more tangential than the development of hydroelectric dams. In the United States, government involvement with scrap aluminum was most conspicuous in two World War II actions. The first was the federal regulation of prices, supply, and demand, a development that placed scrap aluminum alongside other prized war resources from scrap iron to milk.

The second was the most public action, the development among government, industry, and volunteer organizations of scrap drives. Scrap drives encouraged citizens to return their old wares to fuel military production. Iron and steel dominated this campaign, since ferrous metals had both widespread applications and a century of abundant uses, but aluminum, usually in the form of kitchen utensils, was also collected.

Scrap dealers had experience salvaging and selling iron, steel, copper, textiles, and rubber. Using established markets, dealers began to trade in secondary aluminum during the war, and evidence of mass purchasing and processing of demobilized military scrap exists in the inventories of the salvage companies. This wartime trade was limited by government regulations to control prices and speed material flows; scrap dealers regularly clashed with the US Office of Price Administration (OPA) and the Office of Price Management (OPM) over charges of price gouging and hoarding.

Once the war ended, many of these restrictions were lifted. The OPA and the OPM were closed by the end of 1946, and scrap dealers found that purchasing obsolete equipment from the armed forces allowed them to produce highly marketable commodities. Since military demand had led to increased aluminum production, demobilized military equipment became a significant source of aluminum scrap after World War II. War Assets Administration surplus auctions held across the country in 1946 sold off approximately 21,000 airplanes for scrap, sales that generated $6,582,146 ($80,004,128 in 2015 dollars), according to WAA administrator Robert M. Littlejohn. Littlejohn estimated that the aircraft originally cost the military $3.9 billion, but the planes had "accomplished their role as war weapons and their value now [was] chiefly in the aluminum alloy[s] and other metals that can be recovered."[5] Littlejohn said aluminum was in heavy demand for the housing program and for the manufacture of civilian goods, estimating that the sales would produce 200,000 pounds of metal for production of these goods.[6]

The airplanes were sold to five bidders representing manufacturing companies and independent scrap firms. Martin Wunderlick of Jefferson City, Mis-

A collection chairman receiving local citizens' contributions of scrap aluminum for the war at his store, c. 1942. Farm Security Administration, Office of War Information Photograph Collection, Library of Congress (LC-USE6-D-010651)

souri, purchased 5,540 planes for $2,780,000 ($33,790,115 in 2015 dollars). The Sherman Machine and Iron Works of Oklahoma City purchased 7,600 planes for $1,168,550 ($14,203,395 in 2015 dollars). Houston's Texas Railway Equipment Company purchased 4,890 planes for $1,817,738 ($22,094,092 in 2015 dollars). The Compressed Steel Company of Denver bought 1,540 planes for $411,275 ($4,998,931 in 2015 dollars). Sharp and Fellows Contracting Company of Los Angeles purchased 1,390 planes for $404,593 ($4,917,713 in 2015 dollars).[7]

The global scope of World War II meant that airplanes downed or stationed across the planet were potential material for enterprising scrap dealers. Secondary aluminum markets boomed in the United States, East Asia, and Western Europe. Describing the international aluminum trade in the 1960s, the National Association of Secondary Material Industries noted that the two most important markets were the London Metal Exchange and the New York Commodity Exchange. The secondary aluminum prices quoted in these two markets affected trade not only in the United States and Europe, but also (once US restrictions were lifted) in Japan as well.[8]

The Postwar Scrap Trade

The capacity to produce aluminum expanded in Europe and North America. In 1941, the German military had begun building an aluminum smelter in Årdal, Norway; after the war, the Norwegian government seized and completed the unfinished facility, and aluminum production began in 1948. By the 1960s, Norway had the largest aluminum industry in Europe.[9]

The salvage trade in aluminum was initially hampered by the relative lack of history between dealers and customers, as well as by the metal's material properties. Aluminum is not magnetic, which prevented scrap processors from employing much of their equipment and required more hand labor to harvest the material. Furthermore, the aluminum alloys used in aircraft production, foil production, and the other applications that grew during and after the war were new enough that dealers did not understand how to add and subtract the alloying metals to produce suitable scrap. Because of the novelty of scrap aluminum, the scrap trade associations and publications provided guidance on the metal.

Charles Lipsett's primer, *Industrial Wastes and Salvage: Their Conservation and Utilization* (1951), is an early example. The veteran trade journalist Lipsett published the *Waste Trade Journal*, which at the time was the longest continually published secondary material trade periodical in the United States. The *Waste Trade Journal*, founded in 1905, reported prices, sales, and news about the scrap and salvage industry. In addition to founding the journal, Lipsett served on the War Industries Board during World War I, advised the War Department and the US Navy on the disposition of surplus materials after that war, then advised the government price and production agencies during and after World War II. He also participated in the two dominant trade associations representing dealers and processors of scrap materials: the Institute of Scrap Iron and Steel, founded in 1928, and the National Association of Waste Material Dealers (NAWMD), founded in 1913.

In the broad expansion of scrap and salvage operations following World War II, Lipsett published his reference book, explaining the state of the art in each material market. Iron and steel accounted for the largest section of the book, but sections on other metals, glass, textiles, rubber, and paper stock made *Industrial Wastes and Salvage* a useful reference for readers interested in any of the materials reported on in the *Waste Trade Journal*.

Industrial Wastes and Salvage's section on aluminum was quite small. Lipsett discussed the complexities of harvesting nonferrous metals, observing that secondary aluminum was not widely traded compared to ferrous metals. Despite efforts to segregate scrap, the "adequate classification of aluminum scrap types is sometimes difficult."[10] Because of that difficulty, scrap processors were less likely to guarantee the good quality of the metal harvested from aluminum scrap than from scrap iron and steel. Lipsett concluded that "aluminum is being utilized in innumerable forms today and the secondary metal serves as an adequate substitute for the primary ingot derived from bauxite, particularly in such applications where the secondary metal does not have to be of the strictest purity."[11]

The expansion of primary aluminum production meant that potential sources of scrap had grown substantially. "In recent years, the principal supply of aluminum scrap has been obtained from old, salvaged warplanes," however the production of other aluminum goods meant that scrap dealers could also acquire old aluminum from prompt scrap, from old automobiles, and "even from kitchen utensils."[12] In 1947, American smelters processed 411,070 short tons of various alloys of aluminum scrap. Of this amount, Lipsett traced a total of 259,915 short tons of recovered secondary aluminum, in ingot form, to specific alloys, with very little pure aluminum (5,105 short tons) salvaged from prompt scrap. "Aluminum-copper ingot comprised the bulk of these recoveries while copper silicon ingot also accounted for a large tonnage. It is noteworthy that all but one or two percent of the secondary ingot[s] recorded is recovered in the form of aluminum alloys." Lipsett noted: "Little aluminum is recovered in unalloyed form from scrap because it is difficult to extract aluminum from aluminum alloys."[13]

Some twenty years ago [1931], the great bulk of secondary aluminum produced by independent remelters was used in the production of sand castings and for deoxidizing steel. In recent years, great quantities of secondary aluminum are being used for die casting as well because the quality of the product has been improved to an extent that meets with the more exacting requirements of this branch of the casting industry.

The fact that in recent years aluminum alloys have been substituted for brass, bronze, steel, or cast iron in the manufacture of castings, likewise indicates the great progress that has been made in the control of the metallurgical properties of the secondary aluminum now being turned out. Whereas formerly secondary aluminum was unfit for rolling or other working, this is no longer the case.[14]

Lipsett made a case for an expanded secondary aluminum industry in the post-war United States. "The smelter output of secondary aluminum and secondary alloys exceeded 24,500,000 pounds in 1947–48 in two different months," a figure that exceeded

the annual [prewar] peacetime use of primary aluminum. In 1947–48 the use of secondary aluminum alloys fell into the following categories: Casting alloys 63.0 percent; steel deoxidizing 22.6 percent; wrought alloys 8.6 percent; and miscellaneous 5.8 percent. In a questionnaire sent out in 1948, 200 aluminum foundries were asked by smelters to indicate their use of secondary aluminum. Of the various plants, 70 percent stated that they utilized secondary ingots in actual production, 15 percent utilized it for pattern making only, and the remaining 15 percent stated that they did not use any secondary material. However, of the 200 smelters, 92 percent stated that they would use the secondary metal in the future.[15]

Lipsett expressed optimism that the quality of scrap aluminum would soon improve, noting in 1951 that "the Aluminum Company of America (Alcoa) recently announced a method which is designed to produce pure aluminum from scrap. In this process, the scrap is subjected to a caustic process which dissolves the aluminum but leaves all other metals untouched. The process has the further advantage in that aluminum alloys are not attacked by the caustic. The perfection of this recovery method will increase the value of scrap as a substitute for bauxite still further, in economic terms and perhaps also pricewise."[16]

Lipsett's prediction of an expanded market for secondary aluminum came to pass. The production of secondary aluminum in the United States totaled 344,837 short tons valued at $97,450,936 in 1947 ($1,035,763,610 in 2015 dollars), while primary recoveries amounted to 571,750 short tons valued at $161,626,000 that year ($1,717,852,450 in 2015 dollars).[17]

Aluminum scrap imports totaled 143.5 million pounds (71,742 short tons) valued at $17,453,000 in 1948 ($171,645,548 in 2015 dollars). In 1949, imports fell to 80.2 million pounds (40,120 short tons) valued at $10,500,000 ($104,570,000 in 2015 dollars). American scrap importers received materials primarily from Canada, the United Kingdom, Germany, and Italy, with Japan joining the major suppliers in 1949. Whereas scrap metal exports are a major part of American trade in the twenty-first century, scrap aluminum exports from the United States were small in the late 1940s. In 1948, 876,000 pounds of scrap aluminum, valued at $77,800 ($765,142 in 2015 dollars), were exported while in 1949 such exports totaled 784,300 pounds, valued at $49,400,000 ($492,000,000 in 2015 dollars).[18] In 1950, secondary aluminum production in the United States amounted to 221,000 metric tons, about a quarter of total domestic aluminum production for the year.[19]

The Korean War and the 1950s

Wartime needs increased secondary aluminum production during the Korean War. The US military developed programs to save and reprocess its scrapped aluminum goods. Assistant Defense Secretary Charles E. Thomas testified that during the Korean War, air force and navy scrap operations primarily focused on obsolete aircraft. Thomas estimated that scrap processing saved the military between $300,000 and $400,000 annually. NAWMD argued that the US military should sell its secondary aluminum to independent scrap dealers, since doing so would save money. NAWMD executive vice president Clinton M. White testified before the Senate Small Business Subcommittee that the air force lost 1.42 cents per pound on the scrap aluminum it collected at its McClellan and Tinker bases and that it should bid the scrap to independent dealers at auction.[20]

The Korean War's end saw a further increase in secondary capacity and processing. Domestic scrap aluminum production in the United States exceeded 376,000 metric tons in 1955, higher than any year during the war. In 1956, American scrap aluminum production rose to 388,000 metric tons.[21] That same year, the American Smelting and Refining Company announced that its Federated Materials division would build an aluminum smelter at Alton, Illinois, that would be "among the largest ever built for the handling of scrap aluminum." The plant, with a capacity of 72 million pounds a year, doubled Federated's output and made American Smelting the world's largest producer of secondary aluminum ingot. The plant, which opened in 1959, included a scrap-buying operation to purchase the salvaged raw material for the smelter.[22] "Its scrap," reported the *New York Times* in 1959, "is obtained in the form of trimmings, castings, turnings and borings from manufacturing industries, residues from die casting and foundry operations, and 'obsolescent scrap,' the trade's term for such items as discarded kitchen utensils, appliance parts, and automotive crank cases."[23] The *Times* went on to note that the scrap "is sorted, crushed, cleaned, refined, alloyed, and cast into ingots. These are shipped to consumers in the automotive, aircraft, fabricating, and other industries throughout the Midwest. Radio-equipped fork lift trucks, high-speed conveyor belts, four forty-ton and one fifteen-ton gas-fired reverberatory furnaces, three ingot-casting machines, and semi-automatic stacking units are features of the Alton plant."[24]

Although the global recession had briefly reduced demand for secondary aluminum in 1957–1958, the market soon surged again. At the 1959 NAWMD annual meeting, members reported that global scrap sales of all metals totaled more than $4 billion a year ($32.5 billion in 2015 dollars), with about a quarter of the amount from aluminum. American dealers estimated that 5 percent of domestic secondary aluminum was exported.[25]

Three years later, NAWMD dealers were optimistic about the market. "Olin Mathieson Chemical Corporation," the *New York Times* reported, "estimates that 2,550,000 tons of aluminum, a record, will be consumed in the United States in 1962. That amount would be up 13 percent from the 2,250,000 tons estimated for 1961, which equaled the 1959 record."[26] Indeed, throughout the 1960s, scrap aluminum production in the United States progressively rose, exceeding 500,000 metric tons for the first time in 1962 and 600,000 metric tons in 1965, growing by over 100,000 metric tons in each of the next two years, and cracking the million-metric-ton mark in 1969.[27]

Alcoa's growth during and after the Korean War relied in part on processing scrap aluminum. The company had expanded capacity by bringing new smelting facilities on line in 1950 and 1952, then in 1956 completed a $54 million expansion at the Davenport works, doubling that facility's capacity. That expansion complete, Alcoa began building a giant new reduction plant in Warrick, Indiana.[28] The 1957–1958 recession and its effects on global aluminum prices postponed the completion of the Warrick plant, but work resumed in 1963.[29]

Recycling also occurred within factories. Stamping, shearing, or otherwise shaping aluminum into its intended components produced trimmings. This material, known as prompt scrap or new scrap, could be either reprocessed without ever leaving the factory or sold on the open market. As the aluminum industry increased capacity, the generation of new scrap increased.

The 1960s

In 1960, NAWMD changed its name to the National Association of Secondary Material Industries (NASMI) to reflect its stance that its members traded in valuable industrial materials, not waste. That year, NASMI's Charles S. Rosenblum estimated that between 16 percent and 20 percent of the total aluminum supply of the United States came from scrap sources, including "industrial activity (new scrap), discarded or obsolete end products (old scrap), and imports. New scrap makes up about 78 percent of the supply; old scrap about 20 percent; and imports (net) about 2 percent."[30]

Scrap returned to production, Rosenblum wrote, through a network of salvage dealers, some of whom combined the volumes of other dealers as large-scale brokers, and some also acted as processors of the scrap. "As brokers, they promote the collection of scrap using their contacts with a large number of individually [*sic*] small and large sources; and they assure obtaining a fair price by maintenance of an open free market for those materials. As processors, they meticulously transform the scrap into a more acceptable form and condition for the ultimate user by grading and physically shaping, e.g., into briquettes, bales, etc." In this market, Rosenblum argued, the dealer is crucial for consistent, dependable trade. The dealer "services the generator, the source of scrap material. In return, he also serves the user of this scrap aluminum who requires that the dealer meet certain specifications, preparation criteria, and volume through accumulation."[31] Rosenblum estimated that producers of wrought and cast products used 20 percent of the scrap sold in 1960 with the remaining 80 percent used by secondary smelters.[32]

The secondary material markets across the industrialized world experienced growth and change in the decade after Lipsett published *Industrial Wastes and Salvage*, prompting him to write an expanded second edition in 1963. Techniques and capacity to process scrap aluminum had expanded so much in the 1950s that Lipsett concluded in the second edition, "Aluminum is . . . among the more abundant sources of scrap metal, both in its own right and in alloys, since it has been applied in literally hundreds of products which return as scrap to be reprocessed after use."[33] Expanded capacity from secondary smelters helped the market for scrap aluminum to more closely resemble the established market for scrap iron and steel. In 1963, Lipsett concluded, "Aluminum scrap is no longer a real vehicle for speculation; it is a commodity tied basically to primary aluminum, a stable commodity. The margins in both must be made from operation."[34]

Lipsett noted that American scrap dealers recovered 416,000 tons of aluminum alloys from scrap in 1961, an increase of 5 percent from 1960 levels. The independent scrap market was crucial; prompt scrap remained steady at 282,500 tons, but the recovery from old scrap had risen 91 percent above 1960 levels to 133,400 tons.[35] The income generated from selling 340,000 tons of processed scrap in 1961 at $174,000,000 ($1,379,300,000 in 2015 dollars) was "computed from an average price of primary aluminum ingot at 25.5 cents per pound."[36] The vast majority of domestic aluminum recycling in 1961 was done by independent businesses, rather than via Alcoa, Reynolds, or other primary metal manufacturers. "Independent secondary smelters used 331,700 tons, or 67 percent, while primary producers used 44,600 tons, or 9 percent. Foundries, fabricators, and other consumers used 24 percent, or a total of 121,800 tons."[37]

A major development in the 12 years since the first edition of Lipsett's book was the collection of aluminum that had been disposed of by consumers and industries. "The rising ratio of old scrap to new scrap represented to some degree the aging of the industry," Lipsett wrote. "In 1958, for example, 80 percent of the purchased scrap was new; many products had not been in service long enough to be discarded and recycled back to the smelter." The abundance of discarded aluminum goods in circulation meant that "in three years old scrap and sweated pig amounted to a much higher ratio."[38]

Lipsett reported that American consumption of aluminum-based scrap had increased 18 percent in 1962 over 1961, and shipments by the independent secondary smelters had increased 29 percent during the same period. Imports of aluminum-based scrap rose 8 percent in 1962 to 6,496 tons, while exports declined 20 percent to 65,534 tons.[39] The international scrap trade also had evolved, with exports to West Germany, Italy, and Japan growing since 1958. According to Lipsett, these nations "welcome scrap imports since their cheaper labor can sort and handle the scrap much more economically than is the case in this country."[40] The new smelters had better control over separating alloying materials and impurities, improving the produced material.

Since the separation of alloys before remelting was as important for reducing impurities in secondary aluminum as it was for reducing impurities in ferrous scrap before the advent of the open-hearth furnace at the turn of the twentieth century, the advances of the 1950s had made scrap aluminum recycling a feasible industry. Increasingly large amounts of post-industrial and post-consumer scrap made their way into smelters in the 1960s to produce material fit for casting.[41]

Since the first edition of *Industrial Wastes and Salvage*, Alcoa's proposed method to use caustic soda to separate aluminum had been implemented. "The perfection of this recovery method," Lipsett reported, "has increased the value of scrap as a substitute for bauxite."[42] Lipsett predicted that if the price of the secondary metal fell low enough, it would be used in the construction of homes, "where its fire-resistant properties will be particularly advantageous."[43]

Alcoa was no doubt pleased that Lipsett's evaluation reflected one of its goals in designing its corporate headquarters. Although the 1957–1958 recession had cut into sales, the market was improving when Lipsett published the second edition of *Industrial Wastes and Salvage*. The vast array of uses for aluminum by the mid-1960s produced both an abundant supply of material for scrap recycling businesses and an abundance of difficulties in successfully recycling the material. Designs of everything from siding to screens to airplanes fused aluminum to other materials, requiring care in sorting and separating aluminum. Scrap aluminum dealers, like rag and ferrous scrap dealers before them, developed classifications of various grades of scrap. This allowed simple descriptions and trading in volume.

In 1967, NASMI identified nine grades of aluminum scrap. The variety indicates both the widespread applications of the metal and the difficulty of harvesting aluminum from products whose manufacturers did not consider the end-of-life consequences of their designs. Number 1 grade scrap would trade at a higher price than number 4 grade scrap, for example.

1. Old sheet aluminum, except the 70-S series.
2. Old cast aluminum, "once it is freed of any foreign metal or contamination such as tar, dirt, etc."[44]
3. Aluminum pistons from automobile engines.
4. Painted aluminum. "Paint presents a consumer with such problems as additional melt loss, heavy drosses and possible lead or titanium content in the resulting metal. Most painted aluminum scrap originates from aluminum awnings and sidings."[45]
5. Aluminum borings and turnings. "In these items, moisture, as well as alloy mix, cause serious concern."[46]
6. Aluminum screens. "This is one of the most difficult grades for a smelter to melt, and it also consists of an alloy with the highest percentage of magnesium. Sorters are cautioned to watch for screens appearing to be aluminum, but which are made of fiberglass, or metals other than aluminum."[47]
7. Venetian blinds. "Unlike most painted aluminum products, venetian blinds are heavily coated with baked enamel and, consequently, ought to be separately packed to avoid down-grading other aluminum scrap. Lately, some blinds have been made of steel and plastic."[48]
8. Oil cans and juice cans. "The all-aluminum oil can [which emerged after World War II] has practically disappeared since the new oil cans are made of cardboard tube with adhesive aluminum coating. The juice cans

require special packing as they usually contain a coating or sediment, and consist of a high magnesium alloy and possible iron contamination."[49]

9. Aluminum radiators with copper tubes and aluminum radiators with brass tubes.[50]

Aluminum's versatility played into the complexities of recycling it. Designs that adhered the metal to other materials made contamination difficult to prevent. Like other metals, heavy sheet was prized above loose wires or screens, and industry professionals worried about the difficulties of salvaging the metal from goods not designed for disassembly. But Lipsett's optimistic assessment of aluminum recycling in the early 1960s was borne out by further expansions in the rest of the decade as the volume of scrap traded increased.

At first, the increase was steady. Domestic secondary scrap production rose from 441,000 metric tons to 529,000 metric tons between 1961 and 1962, with roughly a third of that old scrap sold on the market. In 1965 the figure rose to 752,000 metric tons. Domestic secondary production cracked the million-metric-ton level in 1969 and hovered around this level for the next five years. The late 1970s saw further increases, exceeding 1.5 million metric tons for the first time in 1978.[51]

The 1970s and Beyond

The major trade association representing secondary aluminum dealers continued to reflect the evolving cultural understanding of waste and environmental responsibility. By 1972, NASMI had once again changed its name. The trade association no longer focused on "secondary materials," instead advocating recycling. NASMI became the National Association of Recycling Industries (NARI). In the twenty-first century, "recycling" remains part of its name; after a merger with the Institute of Scrap Iron and Steel in 1987, the organization is now the Institute of Scrap Recycling Industries.

The name changed, but its operations remained largely the same. In 1973 NARI's grades of scrap aluminum were adjusted slightly to include aluminum cans as number 8, replacing the obsolete oil cans. The energy savings of using secondary aluminum remained, even given the labor and time required to salvage aluminum free of contaminants.[52]

One alloy used in aviation has posed special problems for scrap dealers, but not so much that it has not been collected, processed, and traded. "Alloy 7075 has more than 12 percent total alloying elements other than aluminum. It has the lowest aluminum content of any alloy with a copper content in excess of 1½ percent, magnesium in excess of 2 percent, and zinc in excess of 5½ percent. Consequently, alloy 7075, mixed with other alloys, is generally an undesirable grade of aluminum for the consumer, since it would be uneconomical to treat and would require sweetening the alloy mix in the melt in order to bring about the desired specification in the consumer's end product." Despite this caveat, 7075 still is "highly desirable for those consumers seeking such specification and price-wise could recover as much or more than other better grades of aluminum."[53]

The contamination concerns NASMI had expressed with aluminum siding in 1967 continued. Popular food packaging included materials that were easily separated, such as the metal in many TV dinner packages. Not so easily salvaged was the aluminum fused with paper stock in some fast-food wrappings or the ubiquitous grafting of aluminum with plastics in the small packets offering single-serve portions of ketchup, mustard, mayonnaise, and relish. The cost and difficulty of harvesting aluminum from such applications exceeded the value of the material; the designs of the packets guaranteed they would wind up as garbage. While packaging innovations rendered some aluminum unrecyclable, growth in the use of aluminum cans in the United States and much of Europe heightened the visibility of aluminum waste in ways that promoted the metal's recyclability. Littering concerns about single-use steel cans had been voiced since the 1950s. Conspicuous aluminum waste due to single-use packaging was evident by the mid-1960s. The rise of automobile culture produced roadside litter, often food containers. Federal attempts to regulate litter and commerce on America's roads included an unsuccessful 1958 proposal by Senator Richard L. Neuberger of Oregon and the Highway Beautification Act signed into law by President Lyndon B. Johnson in 1965. In his remarks at the signing, Johnson exclaimed, "There is more to America than raw industrial might," and this act to limit roadside commercial development and waste would begin the process of bringing "the wonders of nature back into our daily lives."[54]

President Johnson had signed the bill after a vigorous public relations campaign by First Lady Ladybird Johnson to beautify roadsides across the country. The work of the president and first lady came after a decade of criticism, not only by writers like Packard, Kouwenhoven, and Blake, but by citizens groups across the country who shared their concerns about blight and litter.[55]

In addition to the Highway Beautification Act, President Johnson signed the Solid Waste Disposal Act of 1965, which established a federal Office of Solid Wastes and provided funds to states and municipalities to plan, develop, and maintain more effective waste management practices. These two acts were part of an era of federal legislation that expanded the regulation of environmental issues over the next 15 years. In 1976, President Gerald Ford signed the Resource Conservation and Recovery Act, which gave the US Environmental Protection Agency (itself created in 1970) the authority to control hazardous waste "from cradle to gave," including the generation, transportation, treatment, storage, and disposal of hazardous waste. Four years later, President Jimmy Carter signed the Comprehensive Environmental Response, Compensation, and Liability Act (CERCLA), more commonly known as Superfund. CERCLA provided broad federal authority to clean up hazardous waste sites and allowed the EPA to identify "potentially responsible parties" that would bear the economic burden of remediation.[56]

Federal interest in matters of waste and pollution worried industries that generated waste and pollution. The existence of a thriving secondary aluminum trade provided disposable packaging producers with a viable method of amplifying their existing efforts to shift responsibility for their trash. Beginning in the late 1960s, KAB's rhetoric kept a focus on the irresponsibility of litterbugs, but also began to champion recycling as an environmental ethic. The public face of KAB was an educational campaign involving billboards, print advertising, and televised public service announcements informing the public that consumers should control the consequences of their waste. In the halls of Congress and state legislatures, KAB lobbied federal and state governments to encourage recycling.[57]

Environmental Recycling

As the scrap aluminum industry cracked the million-metric-ton mark in 1969, the momentum to use recycling as an environmentally responsible approach to consumption and waste increased. The salvage campaigns during and after World War II evolved into eco-friendly recycling campaigns, culminating in KAB's 1970s advertising. The industry's efforts increased awareness among designers that salvaged aluminum was both durable and economically more affordable than virgin aluminum.[58] By the end of the 1960s, collection centers had opened to process aluminum cans. In 1969, Gladwin Hill of the *New York Times* reported, "A small army of overnight conservationists [were] hatched in Los Angeles by a can manufacturer's recent offer of a half-cent bounty on old aluminum cans. The program was begun last May as a pilot program, and it may be expanded to other cities if it works out here."[59]

It did. Municipalities on the Pacific coast and in the Northeast began similar collection programs. In Oregon, a bottle bill requiring the beverage industry to impose a five-cent deposit on beer and soft-drink containers and a two-cent deposit on liquor bottles passed in 1971 despite vociferous opposition from beverage companies. This state-mandated effort to establish producers' responsibility for single-use packaging led to several other states enacting deposit laws over the next 15 years. Manufacturers fought these efforts, and the spread of deposit laws in the United States stopped after 1986. However, the beverage industry supported public policies that did not require producer responsibility. Between 1970 and 1990, more than 10,000 municipalities across the United States established some sort of recycling collection program.[60] Much of the secondary aluminum was used to fashion new soda and beer cans, an activity that at best can be described as static in value and, as McDonough and Braungart noted, risks degrading the metal and creating pollutants.

KAB's marketing dovetailed nicely with NAWMD/NASMI/NARI's evolving rhetoric on the environmental benefits of its members' activities. Since the late 1960s, aluminum scrap dealers had attempted to lobby the federal government, work with regulators, and work with customers to allow them to salvage and process as much scrap as possible. In 1970, the Battelle Memorial Institute of Columbus, Ohio, undertook a research program for NASMI funded by a grant from the Office of Solid Wastes. The ensuing report, published in 1972 by the EPA, focused on the opportunities and challenges facing the various secondary material industries. The study identified three high-priority problems facing aluminum scrap dealers: reclamation of scrap from packaging such as cans, reclamation of scrap from transportation sources, and air pollution control.[61]

In 1972, the reasons to promote aluminum recycling included "improvement of the environment in which we live, and increased need for conservation of natural resources. No longer is economic gain the sole driving force for recycling of waste materials. Social gain has been added in the form of improved living conditions and preservation of resources for future generations."[62] The recycling rate in the United States for aluminum cans in 1970 was under 2 percent, spurring recommendations for more extensive public and private collection efforts.[63] To address problems of contamination from scrap processing as well as to increase the collection of old scrap, the Battelle report recommended that government and industry work together to develop practices and policies to maximize recycling. Suggestions included expanding can reclamation programs, developing laws to segregate scrap at the source, and developing economic systems for municipal recycling collection programs. The report also advocated that the industry "push passage of realistic Federal air pollution laws," an admission of some of the environmental problems facing scrap recyclers.[64]

Alcoa versus Recycling

The secondary aluminum market expanded despite slow movement on recycling by the single largest producer in the industry. Alcoa corporate historians Margaret Graham and Bettye Pruitt noted that Alcoa chairman Irving "Chief" Wilson advocated in the late 1940s and early 1950s for design for recycling, putting less emphasis on aviation-grade alloys in favor of "more practical compositions that could utilize scrap metal, because it could be purchased easily and worked easily by Alcoa's customers."[65] Despite this recommendation, Alcoa's research and development did not prioritize recycling until after the practice had become associated with environmentally moral activities. Alcoa's laboratories then emphasized efforts to recycle its metals, forming a subcommittee on environmental control in 1971 and making recycling one of the largest budget categories of research and development from 1975 onward.[66]

In the early 1980s, Alcoa's giant Warrick, Indiana, plant was reconfigured to emphasize quickly recycling aluminum cans back into new cans and to compete with Reynolds's program to use aluminum scrap for die casting.[67] According to Graham and Pruitt:

> Recycling was barely tolerated within much of the Alcoa community from 1978 to 1982, but because of that it had the advantage of proceeding at its own unhurried pace. When, in 1982, the company faced the radical realization that market conditions would no longer justify further investment in smelters, recycling aluminum offered an economical and publicly popular way of increasing capacity and market penetration. The program was also one of the first credible signs that the Laboratories might be able once again to offer its own independent technical vision.[68]

After 1983, Alcoa's laboratories emphasized recycling programs at the recommendation of statistician and technical planner Charles P. Yohn, who argued that recycling offered a higher return on investment than did process technologies. Four decades after Wilson's advocacy, Alcoa had joined Reynolds and Kaiser at the forefront of secondary aluminum production in the United States.[69]

With all of the major American aluminum companies emphasizing recycling, the volume of secondary aluminum harvested and processed worldwide grew. NARI published a short guide to aluminum recycling in 1981 that brought the history Lipsett had recorded for the 1950s and early 1960s into the era of recycling as an environmental ethic. NARI discussed how scrap processors and smelters acquired and used secondary aluminum in a "phenomenal" expansion between 1960 and 1980. "Tonnage-wise, it has been a remarkable story of the metal's successful ability to reach levels formerly only occupied by copper base scrap."[70]

NARI identified the widely traded forms of aluminum as the 1100 series (unalloyed, basically pure aluminum); the 2000 series (copper as the primary alloying element); the 3000 series (manganese as the primary alloying element); the 4000 series (containing silicon); the 5000 series (magnesium as the primary alloying element); the 6000 series (containing magnesium-chromium); and the 7000 series (zinc as the primary alloying element).[71]

The harvesting of aluminum had expanded in the 1960s, growing as the industry processed more material from industrial manufacturers, consumers, and municipal collection programs. NARI estimated that modern aluminum wire chopping started around 1960 and mass investment in shredders occurred in the late 1970s, some time after ferrous scrap dealers had successfully used shredders to separate metal scrap from automobile bodies.[72]

The major change NARI observed from the market Lipsett described in 1963 was the ubiquity of the aluminum beverage can as a commodity to recycle. "When the aluminum beverage can first made its appearance in the early 1960s, it set the stage for one of the most dramatic consumer recycling success stories in the relatively short history of the aluminum industry."[73] By 1981, more than 2,500 collection programs existed in the United States. Aluminum can recycling data were first recorded in 1970. That year, about 8 million pounds of cans were recycled. Five years later, the weight had grown to 180 million pounds. In 1980, the figure was 600 million pounds, "about 18 percent of total aluminum scrap consumption."[74]

The message sent to American consumers was that it was their environmental responsibility to keep aluminum cans out of the garbage, off the roads, and in the recycling stream. By 1992, these cans were both the public face of recycling programs and a substantial portion of the secondary aluminum market. After 1980, secondary aluminum processing outpaced primary aluminum production, which had started to decline in the late 1970s. Municipal recycling collections supplemented industrial secondary aluminum trading in increasing volumes after 1980, accounting for about a third of total domestic aluminum production during the following decade. Old scrap became an increasingly large source of the total secondary supply, accounting for between one-third and one-half of secondary pro-

Pile of shredded aluminum in front of an automobile shredder, Jackson, MS, 1972. Photograph by Bill Shrout for US Environmental Protection Agency (412-DA-3769). Courtesy National Archives

duction in each year of the 1980s. Secondary production exceeded 1.6 million metric tons every year after 1980 and exceeded 2 million metric tons for the first time in 1988. That year, the supply of old scrap bought and sold on the market exceeded the million-metric-ton mark for the first time, about equal to the new scrap generated. In 1990, old scrap exceeded new scrap for the first time.[75]

In his book *The Evolution of Useful Things*, Henry Petroski noted that the recycling rate for aluminum cans in the United States rose from about 25 percent in 1975 to 60 percent in 1990. In the twenty-first century, Pepsi's desire for secondary aluminum was such that the company offered recycling receptacles to the Chicago Park District on the conditions that Pepsi could advertise on the receptacles and collect whatever cans and bottles were deposited in them. With cooperation from Alcoa, Reynolds, Kaiser, smaller producers, and KAB members, can recycling infrastructure represented one of the most successful material reclamation programs in the industrialized world. In 1992, can recycling programs were able to recycle the aluminum from a collected can into a new can in six weeks.[76]

Environmental Consequences of Aluminum Recycling

The most public face of aluminum recycling involves making new cans from old, but scrap aluminum ranging from dismantled aircraft to kitchenware has been a significant fraction of all aluminum processing for more than half a century. Secondary aluminum processing reduces the environmental damage caused

James Meyer stacks aluminum bricks that have been processed through the densifier at the San Diego Naval Air Station's recycling center, 1993. Photograph by Steve Orr (330-CFD-DN-ST-93-04503). Courtesy National Archives

by primary aluminum production, but recycling brings its own environmental problems. Since the 1950s, secondary aluminum producers have grappled with pollution of the land, air, and water, as well as with the consequences to human health. The US Centers for Disease Control identifies skin irritation, pulmonary fibrosis, and other respiratory system diseases as occupational health hazards associated with aluminum exposure, and regulators in the United States and Europe have observed environmental problems associated with recycling the metal since the 1960s.[77]

A 1970 legal study noted the problem of solid wastes produced by removing impurities in scrap aluminum, wastes that necessitated using collection methods such as a baghouse or electrostatic precipitator.[78] Chlorine gases are frequently used in aluminum production, creating noxious fumes.[79] Burning off impurities in the processing of scrap aluminum may release airborne dioxins and furans, compounds known to be carcinogenic. Liquid ammonia is also a common by-product. These emissions both endanger human health and make processors subject to regulation in the United States and the European Union.[80]

A 1979 British reference manual on waste management noted that pollution control in scrap aluminum processing was a serious problem. Wastes produced by aluminum recycling included fumes from the degreasing and de-oiling of swarf (metal filings) as well as solid wastes (slags and mineral impurities). Processors could manage wastes with investments in technology, but "the severity of the pollution prevention requirements is illustrated by the fact that the cost of pollution control can be about 75 percent of the cost of the basic smelting plant."[81] More than 20 years later, another British survey of the scrap aluminum industry described the salt slag and other waste products from secondary smelting and dross processing as presenting "a problem of disposal. In some areas it is classified as a hazardous waste product and must be dumped under controlled conditions. The disposal cost of slag in Europe is commonly $50 per tonne or more, and this cost and the regulation of disposal has raised interest in the recycling of salt slag."[82]

The industry did invest in technologies to resolve or minimize wastes produced by recycling. Smelters by 1973 used dryer afterburners to consume combustible black smoke. Dust collectors captured particulate matter produced from scrap crushers. Wet Venturi fume scrubbers removed submicron dense white fumes generated during chlorination to remove alloyed magnesium. Furnace well hoods and afterburners collected combustible fumes from oil, moisture, and contaminants, preventing black fumes from entering the atmosphere.[83]

Wet Venturi fume scrubbers use a caustic solution spray to break down the fine particles and dissolve the chemical gases. Once this is accomplished, the solution is scrubbed away and requires its own disposal as a toxic waste. Wet Venturi scrubbers are also "very expensive to buy and install, are difficult to maintain, and the costs are very high to operate, causing additional substantial costs to the smelting operation."[84]

Subsequent technological solutions included means to process dross, salt cake, and ammonia. Despite these efforts, the environmental consequences of recycling aluminum endure. The Toxics Release Inventory reveals that the process of secondary smelting and alloying aluminum produces several undesirable releases. In 2013 reports from New York, Pennsylvania, and New Jersey, secondary aluminum producers acknowledged releases of ammonia, sulfuric acid, hydrochloric acid, chromium compounds, ethylbenzene, hydrogen fluoride, and lead, among other toxins.[85] The closed loop of recycling does result in energy savings and extensions of the life of materials. Producing those technical nutrients, however, releases toxins that could poison workers, neighboring residents, and ecosystems.

Although research into the recycling and potential reuse of the wastes produced by aluminum recycling is a goal of contemporary industrial ecology research (applied in facilities in Europe, Asia, and the United States) and although the contaminants released by recycling pale compared to the ecological damage of mining and smelting primary aluminum, the waste products of scrap recycling must be noted when considering the consequences of returning the metal to production.[86]

A Recyclable Commodity

As the twenty-first century began, aluminum had been widely reclaimed from old industrial and consumer uses for half a century. Today, aluminum recycling furnaces come in three types. Reverberatory furnaces are used to melt a narrow range of feedstock, for example, scrap with a known composition, by passing a hot stream of combustible gases over the aluminum. Rotary furnaces are used to melt a wider range of scrap feedstock. Induction furnaces, which use electricity rather than gas, are used for very clean scrap and produce a much smaller volume of aluminum today than the first two types.[87]

Statistics from the European Union and the United States indicated that aluminum and steel were the two most recycled materials in industrial economies. By 1994, American secondary aluminum production had equaled domestic primary aluminum production; since 2001, secondary material has represented the majority of American production every year. Old scrap represents between one-third and one-half of secondary production.[88]

These statistics characterize the relationship between states and the secondary material industry. Government intervention in the scrap aluminum market emerged again in the late twentieth century. American attempts to legislate producers' responsibility led to deposit laws in a few states. More broadly, attention to recycling as an environmental ethic led the federal government to pass solid waste legislation in the 1960s that encouraged municipalities to study and make more efficient their waste management programs. What resulted was a patchwork system of municipal drop-off and collection programs, more than 10,000 across the United States by 1990.[89]

European approaches have included somewhat more centralized federal programs and guidelines within the European Union on reducing solid waste and returning valuable materials to industrial production. The EU keeps statistics on national recycling rates of various materials, allowing for comparison of different national programs. Federal and international programs cooperate with industry, providing feed for further production. In Western Europe, aluminum recycling

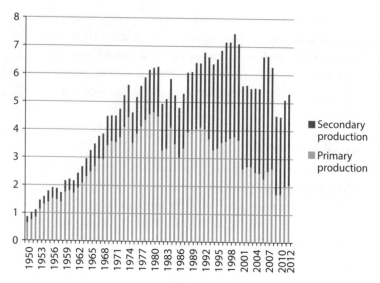

US Primary and Secondary Aluminum Production, 1950–2012 (in million metric tons). Derived from US Geological Survey, "Aluminum Statistics," in T. D. Kelly and G. R. Matos, comps., *Historical Statistics for Mineral and Material Commodities in the United States: U.S. Geological Survey Data*, ser. 140. http://minerals.usgs.gov/minerals /pubs/historical-statistics/ds140-alumi.xlsx (accessed 21 December 2015)

more than tripled in 24 years, expanding from 1.4 million tons in 1980 to 4.7 million tons in 2004.[90]

In 2008, the European Aluminium Association estimated that more than half of the aluminum produced in the European Union that year originated from recycled material, a trend that was on the increase. "In view of growing end-use demand and a lack of sufficient domestic primary aluminium production in this part of the world, Europe has a huge stake in maximising the collection of all available aluminium, and developing the most resource-efficient scrap treatments and melting processes."[91]

The largest aluminum producers in the world happily welcome these collections of post-industrial and post-consumer metal into their operations. Since 2005, the largest consumer of aluminum cans in the world has been Novelis. Novelis spun off that year from Alcan, the largest aluminum producer in Canada and a company in business since 1916. Although the company is headquartered in Atlanta, since 2007 it has been part of Mumbai, India's Aditya Birla Group. (That same year, Rio Tinto acquired Alcan to form Rio Tinto Alcan.) Novelis has operations in the United States, the United Kingdom, South America, East Asia, and Europe.[92]

The largest supply of secondary aluminum purchased by Novelis is used beverage cans. In 2006, the company recycled 38 billion aluminum cans worldwide, accounting for more than 550,000 tons of aluminum and consuming about 45 percent of the used beverage cans recovered in the United States and Canada.[93] Collections from curbside recycling, such as the material sorted at the Sims facility in Sunset Park, get bundled and sold to Novelis. Scrap from construction, demolition, and post-industrial uses, like that produced by the small nonferrous metals businesses in Greenpoint, gets combined by brokers into larger shipments, which are purchased by Novelis, Alcoa, Reynolds, and their global competitors. Unlike most industries, the scrap trade has many suppliers providing material to a small number of large customers.

Novelis purchases secondary aluminum from brokers collecting beverage cans; materials from buildings and construction and from the automotive and transportation industries; electrical systems; lithographic sheets; and foil and packaging. The company's sustainability statement in 2015 claimed: "In order to stay on track to achieve our target of 80 percent recycled content by 2020, we aim to greatly increase the amount of scrap we purchase. Novelis buys both new (from alloy and aluminum production) and old (post-consumer) aluminum scrap from a variety of sources and markets. Currently, the types of scrap we purchase consist of 1XXX, 3XXX, 5XXX, 6XXX, and 8XXX series alloy scraps."[94]

Novelis partners with large manufacturers to develop new aluminum products, and it supplies aluminum sheet and foil to the automotive and transportation, beverage and food packaging, construction, and printing industries. Its history indicates the success private industry has had in returning post-consumer and post-industrial aluminum to industrial production.[95]

In 2014, Novelis opened a $260 million plant in Nachterstedt, Germany, to produce materials for its "evercan," a beverage container made of 90 percent recycled aluminum. Scrap purchased in the United States goes to this facility and to others in Brazil and North America.[96] Zero waste programs in Europe, the Americas, and Asia now encourage recycling as a strategy to reduce solid waste; these collection programs represent the efforts of dozens of nations and thousands of municipalities in the large technological system of aluminum production.[97]

This system works because of the materiality of the metal. For example, alloy 7074 made by Sumitomo, Reynolds, or Alcoa has been grist for both trading and processing. Aluminum's malleability has allowed for secondary trading and processing at a large scale. (It also makes it, like other metals, grist for theft since tracing stolen aluminum bleachers or siding, once it has been shredded and re-melted, is difficult.)

By 1960, the large technological system that produced aluminum had incor-porated an additional industrial loop. The use of secondary material reduced the damage of mining and lowered expenditures on energy. Recycling aluminum allowed more goods to be made at lower cost, providing market incentives for sal-vage. The governmental interventions of the late twentieth century were an aug-mentation of existing scrap aluminum programs and helped to make the metal a material of mass consumption. Producing secondary aluminum is an affordable alternative to mining more bauxite, and the abundance of aluminum to salvage from cans, old aircraft, packaging, and the sundry other uses developed during and after World War II means this affordable alternative is available at a scale large enough to make it a significant input in industrial production.

Between post-consumer and post-industrial sources, secondary aluminum became a majority of the aluminum produced in the United States each year beginning in the 1950s. As already mentioned, salvaging secondary aluminum provides substantial economic savings to producers because the energy use is far less than primary aluminum production requires. Although processing second-ary metal has environmental consequences, they are much less intense than the impact of mining virgin bauxite. For these reasons, secondary aluminum now has cachet as a sustainable material. The US Green Building Council considers use of recycled materials in its Leadership in Energy and Environmental Design rat-ing system of architectural sustainability; although William McDonough did not build Oberlin College's Adam J. Lewis Center with LEED specifications in mind, his use of recycled aluminum for the building reflects that ideal. Recyclability is why sustainable architecture employs aluminum despite the destructive toll of the metal's primary production.

The economics of recycling aluminum have made reclamation a success re-marked upon by twenty-first-century industrial ecologists. K. J. Martchek esti-mated that 73 percent of all aluminum produced globally since 1950 "was still in service" in 2003, and Wei-Qiang Chen estimated in 2013 that more than two-thirds of all the aluminum that had entered the United States since 1900 was still in use. Moreover, secondary material has represented a significant fraction of all aluminum produced for most of its history. By 1950, scrap comprised about one-third of all aluminum used in production in the United States; in 2009 the proportion of scrap in aluminum production was more than 60 percent. If any mass-produced material can claim to have circular material flows over time, aluminum is that material.[98]

The cultural dimensions of aluminum reuse include the rhetorical pleas for efficiency and patriotism in World War II salvage campaigns and the championing of recycling by environmentalists and industries since the late 1960s. By that time, the aluminum industry's promotional efforts had increased awareness among designers that salvaged aluminum was both durable and more affordable than virgin aluminum. Recycling has provided a significant source of American aluminum beyond that used in cans, foil, or warplanes. The variety of goods fashioned from aluminum has expanded since the mid-1950s. Recycled material has made aluminum more affordable and abundant, and any consideration of the goods made of aluminum since 1950 must take into account that recycling helped make those goods possible. The cans, lawn furniture, siding, and wiring discussed in chapter 2 all benefited from recycled material. Goods that Vance Packard did not classify as disposable, including vehicles, durable furniture, and musical instruments, also benefited from recycled material. The second half of this book investigates the history of some of the durable goods produced with aluminum, and how their designs and uses illuminate concepts in sustainable design.

PART II / Designing Upcycled Goods

Chapter Four

Metal in Motion

Aluminum's utility in aviation provided the basis for much of the innovation in aluminum production in the first half of the twentieth century, and aviation remained central in the volume and variety of aluminum produced after World War II. Cold War military investments in aviation by the United States and the Soviet Union were joined by expanded commercial use of the metal as newer, faster jet airplanes transformed large and small airplane manufacturers alike. Military transportation inspired the mass production of aluminum in World War II, and the notion of aluminum as a military material continues to shape rhetoric about the metal more than 70 years after the war ended. In early 2015, the Ford Motor Company began to describe the redesign of its signature F-150 pickup truck as featuring a "high-strength, military-grade aluminum-alloy body."[1]

Although "military-grade" made it into the truck's promotional materials, Ford's intention with the redesign was not to render it suitable for combat operations, but rather to tap into aspects of the environmental ethos associated with aluminum. The basis for that ethos in 2015 had to do with properties of the metal, which made it attractive in the early aviation industry and in the twenty-first century.

Transportation remained a major portion of aluminum consumption in the second half of the twentieth century, with military investments joined by growing commercial uses. The successful scrapping of vehicular aluminum made a closed loop possible. In September 2014, Chaz Miller estimated that while 26.5

percent of the aluminum used in the United States goes into packaging, close behind are transportation products, "which use 23.7 percent. Aluminum is the second-most-used material in new automobiles worldwide."[2]

Between 1946 and the end of the twentieth century, aluminum became a structural material in jet airplanes, bicycles, and automobiles in ways that partially reflect current definitions of upcycling. Because aluminum is light and durable, it has been a staple of aviation technology since the beginning. The cylinder block of the engine that powered Orville and Wilbur Wright's first biplane in 1903 was a one-piece casting in an aluminum alloy containing 8 percent copper. The rapid development of airplanes relied on aluminum, with propeller blades made of the metal emerging in 1907, followed over the next decade by aluminum covers, seats, cowlings, and cast brackets. Aluminum parts were common in the aircraft engines of the next four decades, and manufacturers in the United States and Germany experimented with employing aluminum in the bodies of aircraft as well as in the engines.[3]

Into the Jet Age

As discussed in chapter 1, military investments spurred advances prior to and during World War II. One innovation that deserves special attention was Sumitomo's development of the 7075 alloy in 1936. Mitsubishi subsequently used the alloy for its A6M Zero fighters in 1940. Variations of the alloy were used in American and German military aircraft during the war, and it became a standard for airframes in both military and civilian aircraft after the war. A 7075 alloy uses zinc as the primary alloying element. The result is a metal with fatigue strength comparable to many steel alloys, yet far lighter and less likely to corrode than steel,[4] although 7075 is one of the more expensive aluminum alloys to produce.

The mass production of aluminum in World War II, along with improved alloy strength, allowed aviation companies to make aluminum bodies the norm. Substituting aluminum for steel allowed the US military to modernize its air fleet with the lighter, stronger material, allowing for faster fighters and larger bombers, such as Boeing's B-52. Wartime demand, as well as increased energy capacity to smelt aluminum, transformed the metal from a niche material to an important resource for industrial production.[5]

Wartime aviation embraced aluminum, and initial research into jets found that using the lighter, stiff metal maximized speed and range. In the postwar period, Boeing and McDonnell Douglas expanded their production of modern aluminum-bodied jets for commercial use. Boeing completed its first jet, the XB-47 bomber prototype, in 1947.

The Boeing 707 was one of the most popular commercial airplanes of the late twentieth century. The jet used the 2024 alloy rather than the lighter 7178 alloy Boeing had used on the military version (the KC-135A). While the military priority was to save weight and increase range, 2024 was less prone than 7178 to cracking over time and thus extended the useful life of the 707.[6] By the end of October 1976, 707s had carried just under 522 million passengers for airlines worldwide.[7] Boeing built 1,010 of these planes between 1958 and 1979; they originally sold for $4.3 million apiece ($35 million in 2015 dollars). Most 707s are no longer in service with commercial airlines in the twenty-first century; many have been scrapped.[8]

Several other aluminum-bodied commercial jets followed. Boeing debuted its massive 747 jumbo jet in September 1969 and, as of June 2015, had delivered 1,541 to market. The first 747-100 cost $24 million to purchase ($150 million in 2015 dollars); new 747s in 2015 cost more than ten times that amount.[9] Most of the commercial aircraft in operation around the world since the 1950s are primarily aluminum; the typical passenger airplane features aluminum as 75-80 percent of its weight.[10]

Aluminum is also used in small aircraft. Cessna's first all-aluminum airplane was the 195, produced from 1947 to 1954. Cessna built 1,180 of the planes, initially selling them for $12,750 ($135,514 in 2015 dollars). First sold for small businesses, the 195 developed a following among hobbyists. More than a decade after the production run ceased, used 195s were sold for $5,000. One pilot was profiled in *Flying Magazine* in 1980 with a plane more than a quarter century old.[11] Cessna retained its use of aluminum for a series of small jets it developed between the 1950s and the end of the century, including the 180, 206, 210, 310, and 500 Citation. Used versions of these airplanes sell today for between $250,000 and $400,000, and although many are more than 40 years old, the planes are useful vehicles for businesses as well as status symbols for wealthy individuals. Although aviation companies also work with light carbon-fiber bodies and space-age metals such as titanium, aluminum has provided the structural material for the jet age from the 1940s into the twenty-first century. Aluminum-bodied airplanes of all sizes remain in use by large airlines, small businesses, and wealthy individuals, and those planes that are retired may provide the scrap material that feeds secondary aluminum production.

Cycling Aluminum

Vehicles for a broader consumer market have also employed aluminum. Bicycle manufacturing produced important innovations in the late nineteenth century. David Hounshell argued that the technical advances in the bicycle industry were important for transportation because "refined armory practice[s] and well-developed stamping techniques provided the technical basis for automobile manufacturing in the early twentieth century."[12] However, most of that influence had to do with steel, while aviation spurred developments in aluminum production. Although bicycle manufacturers worked with aluminum, the metal did not provide a significant competitive advantage over hollow-tube steel for frame construction until late in the twentieth century.

Bicycle manufacturers experimented with aluminum in frames and in components as soon as the Hall-Héroult process began to reduce the cost of the metal. Bicycle designers have used aluminum on and off since the late nineteenth century, though the mass production of aluminum-frame bikes did not happen until after World War II, with substantial expansion in the late 1970s. Most uses of aluminum before 1980 involved component parts rather than frame construction. In the decade after General Electric engineer Elihu Thomson made lightweight, hollow-tube-steel-frame bicycles possible with electric resistance welding in 1886, handlebars and rims made of aluminum to reduce weight and resist corrosion were frequently included, and they were commonplace by the 1930s.[13]

A few manufacturers attempted to make aluminum-frame bicycles in the 1890s. Historians Tony Hadland and Hans-Erhard Lessing noted that the St. Louis Refrigerator and Wooden Gutter Company showed aluminum bicycles in New York in 1895, including a racing model weighing 16 pounds. Three years later, the British firm Humber introduced a 22-pound diamond-frame bike built of aluminum tubes mechanically clamped together with steel lugs. According to Hadland and Lessing, the design required inserting steel liners in the ends of the frame tubes, and the resulting frame was not durable. Neither the Humber nor the St. Louis Refrigerator and Wooden Gutter Company designs were produced in significant numbers.[14]

During the 1930s bicycle boom (American manufacturers produced more than a million bicycles in a year for the first time in 1936), Monark sold an aluminum alloy model called the Silver King.[15] A few European manufacturers attempted aluminum frames in the 1930s and 1940s, with most designs only reaching the prototype stage. In 1946, the Council of Industrial Design's "Britain Can Make It"

1948 Monark Silver King. Courtesy Classic Cycle, Bainbridge Island, WA

exhibition at London's Victoria and Albert Museum included automotive engineer Benjamin Bowden's prototype of an aluminum shaft-drive bicycle. It was not put into production.[16] Two years later, the British firms Holdsworth and Hobbs each made protoypes of all-welded aluminum alloy frames. Again, these prototypes did not lead to production models. A year later, Raleigh did the same with a 16-pound bicycle.[17] In Germany, the Hercules HK, introduced around 1958, featured a cast aluminum cross frame.[18]

The British Aluminium company, owned, like Raleigh, by Tube Investments, conducted a study in 1967 for Raleigh's benefit. British Aluminium found that existing aluminum racing bicycle frames were based on the conventional triangulated tubular steel design, which did not take advantage of aluminum's properties to make a lighter frame. Armed with this information, Raleigh built an experimental aluminum monocoque prototype, but did not produce it commercially.[19]

The West German company Heinz Kettler began producing tungsten inert gas (TIG) welded aluminum city and trekking bicycles in 1977 and continued to sell them into the twenty-first century.[20] Kettler claimed in marketing materials that its was the world's first aluminum bicycle, and this claim was cited in a 2015 *Reuters* article about the firm's insolvency.[21]

Bicycle historians identify the end of the 1970s as a turning point in the use of aluminum. For much of the history of these vehicles from the 1880s, steel tubing was the dominant material for racing frames.[22] In the late 1970s, according to historian David V. Herlihy, MIT graduate Gary Klein introduced an aluminum alloy racing frame "that was lighter and more flex-resistant than the conventional steel variety."[23] Shortly thereafter, the Cannondale Corporation introduced several styles of "aluminum for the masses" frames for road and off-road, including in 1983 the Cannondale ST-500 touring bike and in 1984 the Cannondale SM-500 mountain bike.[24] By the end of the decade, Trek was also producing aluminum-frame mountain bikes.[25]

The SM-500 retailed for $595 ($1,357 in 2015 dollars). Six years after the ST-500 made its debut, used versions sold for $500 ($955 in 2015 dollars).[26] More than a quarter century after their creation, Cannondales built in the 1980s are still used, sold, and discussed in forums such as the Adventure Cycling Association Forum. Posts there reveal that some vintage bicycles sell for half of their original list price, reflecting depreciation.[27]

Following Cannondale's success, Raleigh's American branch started to introduce aluminum-frame road bikes. The Italian designers Fabrizio Carola and Carla Matessi developed the Aluetta commuter bicycle with a frame "made up of two half-shells of honeycomb-core aluminum."[28]

Production of aluminum frame bicycles spread to East Asia in the early 1990s, with several manufacturers making TIG-welded frames similar to the ones Cannondale and Klein had popularized a decade earlier. Hadland and Lessing were critical of this development: "Most TIG welded aluminum joints looked a little crude, but, as with steel, improving the appearance was a manual job that cost time and money." They argued that the rise in production of cheap bicycles in Asia "hastened the demise of frame making in North America and Europe."[29]

In the twenty-first century, aluminum remains in regular use as a material for bicycle frames, although several racing bikes now use the even lighter (but more expensive) carbon fiber. Cannondale uses both, continuing to introduce new aluminum designs, such as the CAAD12 racing bicycle.[30]

Automobiles

In addition to its prevalence in aviation and its growing use in cycling, aluminum found its way into the bodies of performance automobiles in small amounts before World War I and then with greater frequency after World War II. The history and present uses of aluminum in automobiles reflect some of the aspirational goals of industrial upcycling, as well as some of the limitations. Geoffrey Davies, the author of *Materials for Automobile Bodies*, noted that a pair of pre–World War I sports cars, the Dürkopp-developed sports car and the Pierce Arrow body (1909), which incorporated rear end panel, roof, firewall, and doors in cast aluminum, were early adopters. However, he concluded, "the Panhard Dyna was probably the first aluminum-bodied car to be mass produced in Europe."[31]

The French engineer Jean Albert Grégoire debuted the Panhard Dyna Aluminium Français Grégoire at the 1946 Paris Motor Show. Grégoire envisioned the aluminum-bodied compact car as an affordable vehicle for French drivers in the wake of postwar austerity. Panhard Dyna mass-produced Grégoire's design as the Dyna X until 1954, when the company replaced it with the Dyna Z.[32]

The expansion of aluminum as a material for automobiles occurred in part due to the efforts of aluminum fabricators to advocate for their metal's use in the 1950s. Robert Cass, the assistant to the president of Cleveland's White Motor Company, published an assessment of metal use in the automobile industry in June 1953. Cass observed a notable change in the field, with less dependence on nickel and a "growing dependence of the industry [on aluminum] and more and more aluminum in the newest models."[33] "The metal showing the largest increased production during the last three years is aluminum." Cass found that many manufacturers were using aluminum to make radiators, since the metal resisted corrosion, and predicted, "Further new uses can be expected in the electrical industry, if only the price of aluminum can be brought into closer relation to [the] cost of competitive copper. In many uses, aluminum displays qualities that meets [sic] standards of car makers and electrical manufacturers alike."[34]

In addition to radiators and wiring, aluminum proved suitable for other components, including some that had environmental benefits. Chemist Eugene J. Houdry, convinced that a link between tailpipe emissions and lung cancer existed, sought to convert the hydrocarbons and carbon monoxide exhaust from car engines into water and carbon dioxide. In 1949, he created the Oxy-Catalyst company to build catalytic converters, ultimately settling on a platinum-impregnated aluminum oxide coating. The device was slow to gain acceptance among American automobile manufacturers, but eventually became standard equipment due to federal regulations.[35]

With the successful incorporation of aluminum in components, one of the major American aluminum producers sought to expand the metal's reach. Reynolds Metals Company published a booklet in 1959 noting reasons for the increased use of aluminum in automobiles and boasting of the inroads its material had made in the industry over the previous decade. "Since the end of World War II," Reynolds declared, "the average amount of aluminum in American automobiles has increased 600 percent," owing to the metal's "light weight, high conductivity, modulus of elasticity, ductility, corrosion resistance, and attractive and varied finishes."[36]

Beyond aluminum's durability, Reynolds claimed that the metal offered economic advantages over steel. "Increasingly, engineers, designers, and purchasing agents have come to realize that they can lower the cost of many parts by switching to aluminum." Reynolds identified most of the savings as coming from two areas. "In the first, aluminum has been used in place of stainless or chrome-plated steel because its purchase price can compete with that of steel on favorable terms. In the second, cost saving has resulted when aluminum's initial price penalty was overcome by directly related cutbacks in production tooling, machining, assembling, handling, shipping, and other cost-hungry operations as well as reducing warranty replacement costs."[37]

In addition to the functional virtues of aluminum, Reynolds touted its aesthetic advantages. "Decorative aluminum always adds to the saleability of a car, because aluminum is universally regarded as a durable, high-quality metal, and because it has appealed to buyers and stylists on [a]esthetic grounds."[38] As evidence, Reynolds showed examples of aluminum decoration on a 1958 Plymouth Fury and a 1957 Chevrolet, concluding: "From the design point of view, one of aluminum's greatest virtues is that it *keeps* its good looks. Unlike plated metals aluminum does not rust, peel, or blister when the protective oxide is disturbed."[39]

Reynolds admitted that its arguments had not yet persuaded the mainstream of automobile manufacturing in 1959, lamenting: "Up to now, aluminum's contribution to lower operating and maintenance costs of American cars has been hidden to a great extent. It has somewhat slowed the increase in car weight, rather than being permitted to lighten it."[40]

The European market was another matter. Several small sports cars made in Italy, the United Kingdom, and West Germany employed aluminum, and the European example could be, the company hoped, a precedent for changes in the

American market. "Booming sales of economical small European cars make it clear that a good many customers want to cut the cost of car operation and ownership."[41] Reynolds cited European research on automobile weight and economic savings to the consumer, noting that André Tranié of Panhard Dyna had determined that the company's all-aluminum Dyna Z1 saved its owner 20 percent of the costs of use over three years even though the purchase price of the aluminum car was 15 percent greater than a comparable steel-bodied vehicle. "Maintenance savings—chiefly in gasoline, tires, and oil—average out in these calculations to roughly *one dollar per hundred miles*."[42] The basic body, including doors and hood, was unusually light, weighing about 215 pounds.[43] Reynolds predicted that some American producers would accelerate their plans to put light cars into production using aluminum.[44]

Reynolds's prediction did not come to pass in the American automotive market of the 1960s. Large steel bodies housing large engines dominated the models General Motors, Ford, and Chrysler built during the decade. One complication for aluminum use was cost, and the example Reynolds offered in 1959 had already been discontinued for five years. After 1954, Panhard Dyna abandoned aluminum bodies for steel due to the cost differences in purchasing and working with the metals. Panhard Dyna had replaced the Dyna X with the Dyna Z1 that year. The Dyna Z1 was short-lived, however; almost immediately it was replaced with the Dyna Z2, which had steel bodywork rather than aluminum. By 1958, only the bumpers, the fuel tank, the engine-cooling shroud, and most of the engine and transaxle cases were aluminum.[45]

Other European automobile manufacturers enjoyed success in integrating aluminum into their designs. Porsche, Aston Martin, and Ferrari used aluminum for the bodies of racing cars during the 1950s and 1960s, enhancing the reputation of the metal with sleek, aerodynamic designs. The British manufacturer Aston Martin produced the DB2/4 between 1953 and 1955 (and the DB2/4 Mark II between 1955 and 1957) with an aluminum body; the car is perhaps most famous for its use in Alfred Hitchcock's film *The Birds*.[46] Collectors still covet these automobiles; a 1956 Aston Martin DB2/4 Mark II was put up for auction at $1.5 million in 2013.[47]

In Italy, Enzo Ferrari had incorporated aluminum into the body of his designs as early as 1940; in the 1950s and 1960s, assisted by skilled aluminum panel beater Alfredo Vignale, he expanded the use of aluminum to reduce weight and increase performance in several models, including the 340. The Ferrari Dino 206 GT, released in 1967, had a body constructed of a combination of aluminum

1956 Aston Martin DB2/4. Photograph by Nick Dimbleby

and steel: the fenders, the hood, and most of the rear were fabricated from aluminum.[48]

Designers found the properties of aluminum challenging, however. Sergio Scaglietti, the designer of Ferrari's GTO and other models starting in the 1950s, explained: "My work was to build and bang the sheet metal. At that time, it was very difficult to work with aluminum. . . . We formed the aluminum over bags of sand. Wood is too hard, you destroy the aluminum, but the sand will move."[49]

This process was too cost prohibitive for mass production, but these essentially handmade vehicles were produced in small runs. Automotive journalist John Lamm estimated that the 36 GTOs made are "among the most treasured automobiles in the world," and today one would not sell for under $10 million.[50]

In 1999, Ferrari produced the Pininfarina-designed 360 Modena with an aluminum body.[51] The 360 Modena was the company's first production car to have its chassis, body shell, and suspension wishbones all made from aluminum.[52] It reached a top speed of 183 miles per hour and listed for $160,000. Ferrari stopped producing the car in 2005.[53]

In the 1970s, Porsche had adopted increased amounts of aluminum in the bodies of its 911 and 928 models in order to reduce the vehicles' weight. In the 911 in 1975, aluminum was 10 percent of the body; in the 928 in 1977, the amount of aluminum had almost doubled to 19 percent of the body. The proportion of ferrous metal between the two models declined from 58 percent to 53 percent. The remaining body weight consisted of the other materials that had become commonplace in automobiles by the 1970s, including glass for windshields, plastic for much of the interior, rubber for tires and belts, and textiles for seating.[54] A 1978 analysis found that aluminum made up 18 percent of the vehicle's weight, breaking down as 39 percent of the engine's weight (aluminum was in the radiator, exhaust system, and air conditioning compressor), 23 percent of the gear box weight, 21 percent of the chassis weight, 10 percent of the overall body weight, and 1 percent of the electrical equipment's weight.[55]

The vehicles using aluminum bodies in the 1950s, 1960s, and 1970s tended to be small-run, high-performance sports cars that were expensive. In 1993, the American Society for Metals' Sohan L. Chawla and Rajeshwar K. Gupta noted that although aluminum resisted corrosion and was lighter, it was not a popular choice for mass-market vehicles. "Until now, aluminum frames and body panels have been limited to low-volume, high-price-tag luxury cars."[56]

In 1990, Honda had unveiled another aluminum sports car. The Honda NSX was touted by Friedrich Ostermann in 1993 as the "only all-aluminum vehicle which is made in a production run. To what extent it can be regarded as a showpiece for aluminum materials technology in vehicle construction is not yet known."[57]

The NSX had the longest production run to date of an aluminum-bodied automobile. The origins of the NSX lay in a commission by Honda to the Italian car designer Pininfarina to design the HP-X (Honda Pininfarina Xperimental) in 1984. Over six years, Honda chief designer Ken Okuyama and executive chief engineer Shigeru Uehara developed the NSX prototype to compete with the Ferrari 328 (later the 348), with the goal of meeting or exceeding the performance of the Ferrari, while offering greater reliability and a lower price point. The NSX was the first production car to feature an all-aluminum monocoque body, incorporating a revolutionary extruded aluminum alloy frame, and an all-aluminum suspension. Honda engineers estimated that using aluminum in the body saved nearly 200 kilograms in weight over the steel equivalent while the suspension saved an additional 20 kilograms.[58] The production car debuted at the 1989 Chicago Auto Show and was available for sale from the summer of 1990 until 2005 as the Honda NSX in most of the world and as the Acura NSX (the luxury-brand name) in the United States.[59]

The NSX's all-aluminum body met with critical acclaim and emulation. In a 1990 review of the Acura NSX, *Popular Science* critic Dan McCosh noted that "by using almost as much aluminum as on a dirigible," the NSX was able to distribute much of its weight in the rear of the car. "With 50 percent of the car's mass on the rear axle, the NS-X is more in the Porsche 911 mode than most modern sports cars," improving performance, agility, and handling.[60] Publishing in 2012, Geoffrey Davies argued that the 15-year production run of the NSX (in all, about 18,000 cars sold worldwide) proved that "an aluminum body built using conventional manufacturing methods . . . was and is possible in moderate numbers."[61]

The NSX list price in the United States ranged from $60,000 in 1991 ($104,412 in 2015 dollars) to just under $90,000 when Honda ceased production in 2005 ($109,224 in 2015 dollars). In 2012, *Car and Driver* estimated that the values of used NSX models in top condition ranged from $25,000 to $33,000 for model years 1991–1994, with values progressively rising to $42,000–$58,000 for model years 2000–2005.[62]

Honda's NSX came to market during an era of environmental concerns about automobiles. Alcoa used the language and methodology of industrial ecology to promote the expanded use of aluminum in vehicle construction. In 1994, it released a life-cycle analysis of aluminum body structures in automobiles. Alcoa concluded that substituting aluminum for steel could "reduce the body weight of a family-sized car by 300 pounds, or 40 percent. This directly results in nearly a 10 percent reduction for the vehicle. If other systems in the car also use aluminum, and if secondary weight reduction opportunities are fully captured, total car weight can be reduced by as much as 25 percent."[63] With these savings, Alcoa argued, an aluminum body structure "becomes more energy efficient than a steel body on a life cycle basis in as little as 14,000 miles of driving, saving the lifetime energy equivalent of 280 gallons of gasoline."[64] These advantages would grow as the use of recycled aluminum became more widespread in automobile manufacturing.

Aluminum Vehicles as Technical Nutrients

The life cycle of vehicles partially reflects McDonough and Braungart's model of a circular material flow. As mentioned above, retired airplanes are an excellent source of scrap aluminum, and some of the great prizes for scrap dealers in the mid-1940s were in the form of retired military aircraft. Since vehicles, according to rubbish theorist Michael Thompson, have a life-span of decreasing value due to mechanical wear, once they are no longer sufficiently functional they can provide technical nutrients to produce new goods. How they do this, however, reflects the complexities of scrap recycling discussed in chapter 3.[65]

These complexities deserve elaboration specific to the history of recycling cars. Iron and steel recycling firms found that salvaging old automobiles accounted for a third of their material by 1980. This harvest was made possible by the proliferation of large shredders, which reduced an automobile to its component materials in minutes. Magnets separated the ferrous scrap from the plastics, rubber, textiles, glass, lead, petroleum, and other materials, which were turned into a pile of fine particulate matter that became known as automobile shredder residue (ASR), or "fluff." The adorable nickname aside, fluff is toxic waste that has subjected scrapyards to civil lawsuits and government regulations. Its creation is due to vehicles being designed with performance in mind, but not with disassembly.

Scrapping large aircraft, such as 707s, is a slower process involving torches and shears, and it does not produce shredder residue. Aluminum-bodied automobiles to date have had a small share of the car market, but they are shredded. During the 1990s, industrial ecologists evaluated the complexities of reclaiming aluminum from these vehicles by either removing components by hand or separating out aluminum from shredded automobile bodies.[66] The Society of Automotive Engineers' Adam J. Gesing and Aron Rosenfeld observed in 1996 that difficulties for recycling included the possibility of different suppliers using different alloy combinations for the same components.[67] "Approaches to dealing with mixed alloy scrap will therefore include alloy rationalization, selective dismantling, and upgrading the scrap by sorting scrap into cast and wrought fractions, [into] major alloy families, or into individual alloys."[68]

Separation by 1996 was made possible by laser optical emission spectroscopy, which provides an accurate piece-by-piece analysis. Gesing and Rosenfeld concluded that although risks of coated and dirty scrap exist, the current sorting methods for aluminum scrap from vehicles in the recycling system were adequate, and technological improvements were available should recyclers require better sorting processes.[69] At the end of the twentieth century, the process of reclaiming aluminum from old vehicles was successful, but it did create wastes that could have consequences to ecological and human health.

"The Most Sustainable Truck"

Environmental concerns can help to shape production. In the wake of the 2008 economic meltdown, the US government offered General Motors and Chrysler a bailout. As it did, the government also raised fuel-efficiency standards. The Ford Motor Company was not part of the bailout, but it adopted aluminum as a way to raise its fleet's fuel efficiency. In 2014, Ford announced that its F-150 truck, the best-selling pickup in the United States for 37 consecutive years (and the product credited with 90 percent of Ford's profits) would now feature aluminum rather than steel in its body.

While the properties of aluminum that Ford values are similar to the ones that were advantageous in the NSX and 707, Ford's language to describe its material choices is rooted in sustainability. Touting the new F-150 as "not only the best performing but the most sustainable truck ever to roll off a Ford assembly line," Ford claimed that by "dramatically expanding the use of high-strength, military-grade aluminum alloy in its body," it had cut the truck's weight by up to 700 pounds. The light weight of the metal was crucial to the truck not because it could then carry bombs in flight, but because it would use less fuel. In addition to touting the weight savings of aluminum, Ford highlighted the recyclability of the metal. "Ford made a big investment in closed-loop recycling for the 2015 F-150, partnering with aluminum suppliers Novelis and Alcoa to recycle aluminum scraps from Ford's manufacturing process directly into aluminum for more F-150s. These scraps, most of which come from stamping windows into body panels, make up as much as 40 percent of the original metal used."[70]

Ford had given chief engineer Pete Reyes the task of developing the new F-150 in February 2010. Reyes enjoyed working with aluminum. "One beautiful thing we learned about aluminum," he said, "was that you could actually cut and weld it. You can't cut and weld the boron and other ultra-high-strength steels in some areas in previous models, so you have to replace the whole section."[71] "And the truck is designed so that sections are easier to repair or replace. We went from a two-piece to a three-piece frame so we could section it easier." Reyes's innovation made it easier for drivers to extend the life of the truck by replacing damaged parts. "The aluminum front end is now simpler to assemble because there is one less piece, it's a stiffer, stronger attachment and much easier to replace if it's damaged in an accident."[72]

Producing these trucks on a large scale required investments to adapt to aluminum. According to vice president of manufacturing Bruce Hettle, Ford spent

The redesigned 2015 Ford F-150 pickup truck represents the most ambitious use of aluminum in a mass-marketed automobile to date. Courtesy Ford Images

three years planning and retrofitting its plants in Dearborn, Michigan, and Claycomo, Missouri, at a cost of more than $1 billion.[73] Ford also helped dealers with the $30,000–$50,000 cost to retrofit their repair shops, and it trained employees both at dealerships and at hundreds of independent repair shops in how to perform maintenance on aluminum.[74]

Aluminum producers are partnering with automobile manufacturers to fabricate more secondary material; the two industries share so much interest and expertise that Novelis hired former Ford executive Phil Martens to be its chief executive in 2009 (but fired him in April 2015).[75] In 2015, Novelis opened a recycling facility in Oswego, New York, capable of converting as much as 10,000 metric tons of recycled scrap into automotive sheet aluminum per month.[76] *Forbes's* Joann Muller described the process of reclaiming aluminum scrap for the F-150:

> When a vehicle body panel is stamped, about 40 percent of the metal winds up as scrap. Instead of gathering up all the various metal scraps from its stamping plants in Dearborn, Mich. and Buffalo, N.Y., Ford installed $60 million worth of elaborate pneumatic scrap-handling equipment that will

separate the aluminum alloy scraps on conveyors and deposit them in dedicated containers to avoid contamination by other grades of metal. Novelis
contracted a fleet of 150 trailers to ship the scrap, in pristine condition, back
to its Oswego plant for reprocessing. Scrap from Alcoa, another supplier, goes
back to its plant in Davenport, Iowa.

The loose, shredded scrap is received in bulk dump trucks at the Novelis
plant and is then dried to remove any moisture or oil. The pieces are then
melted in a 2,000-degree furnace, with extra ingredients added to rebalance
the specialized alloys. Once the molten metal is ready, remaining impurities
are removed and it is cast into massive 30,000-pound ingots for subsequent
processing. It's then ready to be rolled into sheets one-sixteenth of an inch
thick and shipped in giant coils back to Ford's stamping plants, where the
process begins anew.[77]

Novelis's vice president and chief sustainability officer, John Gardner, declared
this to be a "truly integrated" partnership with Ford that "has enabled us to collaborate and build an infrastructure that is ensuring Ford's automotive aluminum
is recycled in a truly closed loop, recreating the same automotive sheet again and
again and again."[78]

Ford gutted its entire Dearborn manufacturing facility to make the switch
from steel to a more sustainable, lightweight aluminum, and Novelis constructed brand-new automotive aluminum production lines and new recycling infrastructure to process the return scrap. We even collaborated on the
design of a unique fleet of trucks to deliver Novelis aluminum to Dearborn
for stamping and pick up the return scrap for closed-loop recycling back at
Novelis.

Novelis and Ford also have been collaborating on the design of the vehicles themselves, using aluminum alloys that accept higher amounts of recycled content and planning with end-of-life recycling in mind.[79]

Early returns on Ford's investment were promising. In November 2014, the
EPA released a fuel-economy rating of 19 miles per gallon city/26 miles per gallon highway, for a combined fuel economy of 22 miles per gallon.[80] Ford's lowest retail price for a 2015 F-150 was $26,030.[81] It sold about 10,000 units in the
truck's first month on the market, more than half the total number of NSX cars
sold worldwide in 15 years.[82] The early success of the aluminum F-150 inspired
emulation; Honda announced plans to revive the NSX in 2015 and began producing a 2017 model at its Marysville, Ohio, plant in May 2016.[83]

End-of-life issues for the aluminum Ford trucks are similar to those of the older steel models. Both versions run the risk of immediate functional obsolescence due to wrecks. Both can be disassembled in a matter of seconds by shredders. Both then produce toxic waste as a result of shredding because of the mix of materials in the vehicle's design. Although Ford has incorporated aluminum for environmental reasons, and although the aluminum can be harvested and continually upcycled, the lack of design for disassembly means that the scrap processors harvesting that aluminum are creating ASR. As a source of technical nutrients, the F-150 retains the toxic complexities of earlier automobile designs.

The F-150 also shares a complicated cultural identity with earlier vehicles. While it is true that airplanes, bicycles, and automobiles may all be rendered functionally obsolete due to sudden or long-term damage, all also are capable of becoming emotionally significant objects to their users. The automotive historian David Lucsko noted that users of older vehicles have formed clubs, published guides, and lobbied to save older vehicles from scrapping. The price at auction of a 50-year-old Aston Martin or Ferrari is high both because of scarcity and because of a cultural valuation of the vintage automobile. This culture, Lucsko argued, is not acknowledged in Michael Thompson's model of how automobiles lose their economic value over time.[84]

Lucsko's argument is intuitive for historians and students of design. Good design should create functionality and connection with the user. Although vehicles may be rendered less efficient due to use, emotional attachments may extend their lives as durable goods either through repair of the working vehicles or as museum pieces on display. This complication is relevant in the consideration of recycled materials as technical nutrients in a closed loop of resources in an industrial economy. Even if the aluminum F-150 becomes the standard of new automobiles, many vehicles will not be scrapped. Other sources of secondary aluminum may be needed to fill the breach in the open loop, or Ford may rely more on primary aluminum (despite the environmental damage primary aluminum brings) to supply the market with new aluminum vehicles.

For the most part, the history of aluminum vehicles is consistent with the model of technical nutrients being recycled into industrial production. A vehicle (be it an airplane, a bicycle, or an automobile) is used until wear or a crash inflicts damage that renders it unfit as a vehicle. The aluminum in it is then scrapped, melted, and returned to industrial production.

Occasionally, however, the experience of individual vehicles deviates from the pattern. The Aston Martin auctioned a half century after its manufacture represents a highly valued vintage good. Although it is in the minority, its story illuminates an open loop in material use. The Aston Martin's body is taken out of the cycle not because it is landfilled, but because it is considered a durable good that retains value.

Lucsko argued, however, that this is not a unique phenomenon. In his work on automobile culture, he discussed the vintage automobile enthusiasts who maintain and modify aging vehicles, often well beyond the recommendations or desires of the manufacturer. These enthusiasts, in Lucsko's words, "spend billions annually on parts and accessories, because whatever their specific interests, theirs is not an economically rational approach to the automobile and its utility. Instead, the car itself is paramount."[85]

The close emotional connection some users have with their vehicles makes such stories frequent. This behavior complicates Michael Thompson's model of vehicle value declining over time, instead making these cars (and bicycles and airplanes) into culturally contingent objects that may increase or decrease in value depending on scarcity, cultural norms among collectors, and wear over time. These cultural relationships are well established for vehicles produced in the 1950s and 1960s. The new Ford F-150 is unusual in the history of aluminum vehicles since it was designed specifically to be sustainable, and it is being produced on a scale far greater than that of other aluminum vehicles. Its experience should be revisited over time to see how users drive, modify, preserve, or scrap individual trucks. It is unknown whether the aluminum F-150 will become both a product of and a source of technical nutrients, or whether it will become primarily a durable good taking materials out of the cycle of production. That distinction is relevant to other goods made of aluminum since the 1940s, as the next two chapters discuss.

Chapter Five

Covetable Aluminum Furniture

Owning an iconic representative of mid-century modernism is easy if you have enough money and an internet connection. A consumer wishing to purchase an Eames Aluminum Group lounge chair or executive chair can go to the Design Within Reach website, search for "Eames chair," and, with a couple of clicks, have a chair delivered. These are impulse purchases for few, since the lounge costs between $1,849 and $2,559 and the executive chair between $3,299 and $3,939.[1]

Why would anyone spend thousands of dollars on an office chair? A comfortable chair is important, but comfort can be had at a lower price. The desire to purchase one of these chairs comes from both its functionality and its identity as a high-status symbol of modern design. Because of this, demand for these chairs remains strong, and they have remained in production for more than half a century.

Herman Miller, Inc., builds the chairs in its Greenhouse building in Holland, Michigan. William McDonough designed the 295,000-square-foot office, manufacturing, and distribution center in 1995 to (as the American Institute of Architects said when it recognized the project as one of its "top ten" winners) maximize occupant comfort, health, and communication; integrate the exterior landscape; and maximize use of daylight in illuminating the facility.[2] In the Greenhouse, Herman Miller workers incorporate aluminum into designs more than half a century old. One could try to find a used Eames chair made 20, 30, or

40 years ago to save money compared to the price of a new one. But those older versions are also expensive.

Prewar Aluminum Furniture and Postwar Opportunities

The past and present of the Eames designs illuminate how aluminum has become part of luxury furniture. Before aluminum became more accessible after World War II, designers in Europe and the United States coveted the material for its malleability. In the 1930s, the French designer Jean Prouvé developed Cité chairs, which combined aluminum and leather. Prouvé sought to design furniture that could be prefabricated, functional, easy to mass-produce, and affordable. His Cité chairs were strong enough to hold the weight of adult humans while being lighter than the tubular steel frames of, for example, Marcel Breuer's Wassily chairs.

Prouvé also worked with steel during the 1930s, but as aluminum prices declined during and after World War II, he expanded the aluminum Cité chair's distribution. His postwar uses of aluminum extended to prefabricated housing; most famously, he developed in the late 1940s the all-aluminum Maison Tropicale houses, which were intended to be affordable, durable shelter for use in Africa.[3]

Lawrence Kocher and Albert Frey's Aluminaire, an aluminum and glass box house built in the international style for the 50th anniversary show of the Architectural League of New York, won sufficient acclaim to be one of two American houses included in the Museum of Modern Art's first architectural exhibition in 1932.[4]

Matters of taste have shaped the markets for aluminum furniture. As historian Clive Edwards observed, metal chairs and desks had been employed mostly in institutional settings prior to 1945. Furnishings for the home were primarily composed of woods and fabrics. A 1931 trade report noted the bias: "However original and striking when seen at exhibitions, [aluminum furniture] may seem incongruous and distasteful to many people in regard to ordinary use in the home."[5]

While designers challenged these norms in the 1930s, taste and economics limited the proliferation of aluminum furniture during that decade. Designers in both the United States and Europe found new opportunities in aluminum once World War II was over, and the metal found its way into buildings and furnishings of all sorts. In 1956, architect Ludwig Mies van der Rohe began his contribution to the Reynolds Metal Company's two-volume survey *Aluminum in Modern Architecture* with a warning: "The danger with aluminum is that you can do with it what you like; that it has no real limitations."[6] This durable, malleable material was both newly abundant and particularly able to fill the breach at a time when material shortages were causing logistical problems on both sides of the Atlantic.

As Clive Edwards noted, a timber shortage in Great Britain combined with an abundance of aluminum to give the metal more consideration in domestic furniture design. The London department store Selfridges displayed an exhibition entitled "Aluminum—From War to Peace" in the summer of 1945, showing a wide range of household goods from chairs to lamps to kitchen appliances constructed from aluminum. The exhibition subsequently traveled around England to positive reviews and inspired other displays of aluminum furniture and kitchenware.[7]

A 1946 article, "The Light Metal Home" in the journal *Light Metals*, advocated for greater use of aluminum in furniture and suggested that bold designers would transcend the "complete lack of any really courageous scheme of design" that would both suit the metal's characteristics and appeal to the general public.[8]

In both the United States and the United Kingdom, aluminum manufacturers worked with designers to produce attractive new uses for the metal. An early success in the United Kingdom was Ernest Race's aluminum BA chair, which was displayed in the 1946 "Britain Can Make It" industrial exhibition. The chair was composed of cast aluminum coated with enamel; the frame was adhered to a plywood seat and back covered by vinyl or leather. The BA chair sold more than a quarter million units between 1945 and 1969. The Esavian school chair became a staple in late 1940s British classrooms and was subsequently imported into American classrooms.[9]

In the United States, aviation companies not only built airplanes for domestic use after the war, but also entered furniture construction. Cessna, a maker of light aircraft, built storage units with aluminum drawers immediately after the war. The trade journal *Modern Metals* reported that these products were popular enough that Cessna tested selling the line at Chicago's Marshall Field's department store, and mass production started at Cessna's Hutchinson, Kansas, factory.[10]

Herman Miller's "Good Modern Design"

The Herman Miller furniture company of Zeeland, Michigan, represents the way furniture production has evolved over the twentieth and early twenty-first centuries. Founded as the Star Furniture Company in 1905, the firm focused on making reproductions of traditional home furniture in its early years. Renamed the Herman Miller Furniture Company in 1923 when president D. J. De Pree convinced his father-in-law, Herman Miller, to become majority owner of the firm, the company changed its emphasis during the Great Depression. Then it began collaborating with designer Gilbert Rohde on innovative products that would reshape the looks of homes, airports, and especially offices in ways that remain recognizable in the twenty-first century. In his memoirs, D. J.'s son (and successor) Hugh De Pree recalled a value he saw when he started working in the family firm in 1938: "From our experience in making both traditional and modern furniture in the same factory, we had learned that in modern we were delivering more furniture per dollar. We were also sure that good modern design would have longer life, therefore becoming an answer to every manufacturer's dream for repetitive cuttings of the same components."[11] Hy Bomberg, who joined Herman Miller in 1950 and became a senior marketing manager, agreed with that emphasis: "Quality meant that you bought things to last. The whole essence of Herman Miller has been the consistency of quality: the ability to design something that would last, not only physically but spiritually, because we're not in the cosmetic design business."[12]

Rohde brought Herman Miller into the office furnishings market in 1942, beginning a direction that would continue after Rohde's death in 1944.[13] De Pree then hired *Fortune* writer George Nelson to be the firm's first design director in 1945.[14] Nelson kept the position until 1972. He supervised an innovative period of creation as Herman Miller manufactured much of what came to be known as mid-century modern furniture in collaboration with the designers Alexander Girard, Isamu Noguchi, Harry Bertoia, Bob Propst, and, most famously, the husband and wife team of Charles and Ray Eames.[15]

Although the name of the company is Herman Miller, Miller's son-in-law and grandson have shaped the firm. Hugh De Pree assumed the management of the company in the late 1950s during his father's illness. He became president and chief operating officer in 1962, and in 1969 he also became chair of the board. In 1970, Herman Miller stock was offered to the public for the first time, and Herman Miller Furniture Company became Herman Miller, Inc.[16] By the time of Hugh De Pree's death in 2002, Herman Miller had developed a catalog that made extensive use of recycled aluminum. How and why it did deserves examination, not least because of the innovations of Charles and Ray Eames.[17]

Making Modernity: The Eames Office and Aluminum

The Eames Office pioneered new uses of plywood, fiberglass, and aluminum in the 1940s and 1950s, arranging, in Charles Eames's words, "elements to accomplish a particular purpose." Eames viewed the work of design as "an expression of purpose" that "may, if it is good enough, be later judged as art."[18]

Wartime production provided affordable supplies of aluminum; it also transferred knowledge and technological innovations to domestic design. Ray Eames observed that wartime material embargoes shaped the way the Eames Office approached design: as a way to address problems. "Materials were not available," she recalled in 1980, "and that's when we started to work on the splint as a way of contributing to the war effort. And doing the production of that we were able to develop techniques that could then be applied to the furniture, which after the war we brought to a point which was then shown at the Museum of Modern Art."[19] Ralph Caplan noted: "They had, in 1942, been commissioned by the Navy to develop lightweight leg splints, which were manufactured by the Evans Products Company in Venice, California. That project gave the Eameses access to molding technology developed by the British for mosquito bombers."[20]

The author of *Classic Herman Miller*, Leslie Piña, argued that the firm became more influential after World War II, when "the really alert designers began to introduce inspired forms of truly functional furniture that even looked original. It was designed from the inside out, and it could be appreciated from the outside in. Plus, it could be mass produced and marketed for huge populations of people in the workplace who suddenly needed furniture to accommodate new ways of doing business and better systems for organizing information."[21]

The materials of Eames furniture fall into four main technological groupings: molded plywood, molded reinforced fiberglass, molded and welded wire, and cast aluminum. The Eameses relied on the expertise of their staff, and in the case of aluminum Don Albinson and Bob Staples were especially important. Staples learned to cast aluminum molds at 910 Washington Boulevard, the home of the Eames Office for more than 30 years, and carved the wooden "antlers"—the supports—for the Aluminum Group chairs himself.[22]

In keeping with aluminum's use in cheap applications, such as folding chairs, the first Eames furniture to use the metal was affordable chairs. For the Museum of Modern Art's 1948 International Competition for Low-Cost Furniture Design (motivated by the urgent need in the immediate postwar period for low-cost housing and furniture adaptable to small dwellings), the Eames Office submitted designs for furniture produced from stamped aluminum or steel. In the text accompanying the designs, Charles wrote: "Metal stamping is the technique synonymous with mass production in this country, yet 'acceptable' furniture in this material is noticeably absent. . . . By using forms that reflect the positive nature of the stamping technique in combination with a surface treatment that cuts down heat transfer, dampens sound, and is pleasant to the touch, we feel that it is possible to free metal furniture from the negative bias from which it has suffered."[23]

The Eames Office failed, however, in making "acceptable" furniture with its first attempt. The company produced prototypes of an upright chair with steel and aluminum seats with intended retail prices between $5.80 and $11.73 ($57 and $115 in 2015 dollars).[24] Such a chair would have been more expensive than the aluminum folding chairs Vance Packard was soon to decry as trash, but they would have been sufficiently affordable to most American consumers. Although Charles Eames hoped that metal stamping would be an economically viable production technique, no Eames chairs were ever mass-produced in this material.[25]

The Aluminum Group

When the Eames Office did succeed in making aluminum furniture for production, the prices of the models were not nearly so affordable. The Aluminum Group pieces, even under their original name of indoor-outdoor furniture, were immediate successes, but also expensive. In 1989, Ray Eames and onetime staffers Marilyn Neuhart and John Neuhart wrote: "The [original Aluminum Group] chair's design required combining factory techniques with expensive hand labor and craftsmanship."[26] Even in its early days, Eames aluminum furniture was not affordable to the masses.

The Aluminum Group began in 1957 with an outdoor chair with a single stem base supporting a sling seat composed of synthetic material padded with foam. The development of the Aluminum Group of furniture (also, at first, often referred to as the Leisure Group) in the Eames Office was, according to Don Albinson, a relatively fast operation, compared to the time and energy that went into the development of the fiberglass chairs and the Eames lounge chair and ottoman. He recalled that the entire process of prototype making, mold developing, materials testing, and casting required only about a year from start to finish. "It was amazing," he said, "that we got the whole line of furniture ready in so short a time."[27] In early 1957, Albinson remembered, Charles had returned to the office from a trip with a new project in mind. He and designer Parke Meek had just spent a few days in Santa Fe, New Mexico, with Alexander Girard photographing the Girard aluminum storage system designed for an Alcoa advertising campaign. Neither the plywood nor the fiberglass furniture had stood up to the tests of prolonged exposure to weather and changes in temperature.[28]

Herman Miller 1959 catalog display of Eames indoor-outdoor furniture. From the Collections of the Henry Ford; gift of Herman Miller, Inc. (object ID: 89.177.951)

Charles Eames credited the origins of the aluminum chair to a discussion he had with Girard. Speaking in 1958, Eames noted:

> This one started when Alexander Girard, Sandro [as Girard was known], came to visit and we were talking about furnishing a house which he and Eero [Saarinen] had just completed. . . . Sandro was bemoaning the fact that there was no real quality outdoor furniture. . . . You start on a close human scale. Here is a friend who has done something. He needs something for it, and you become involved. As we were trying to analyze the reasons why there

was nothing available on the market to suit him, why we were of course starting to write a program for designing the object to fill this void. . . . This was
not like the beginnings—or even the motives—of the other chairs. The story
of those was mostly of sticking to a concept. . . . This was more like an approach to an architectural problem, where you have the program fairly well
embedded and call on past experience.[29]

Problem solving was at the center of Eames's conception of design, and the
early Aluminum Group and the subsequent furniture the Eames Office designed
using aluminum reflect how this modern metal solved existing problems. In this,
Charles was aided not only by the expertise in the office, but by incentives from
Alcoa and Herman Miller. In 1959, Alcoa devoted $3 million to encourage alumi-
num design, and at the same time Herman Miller worked with the Eameses to
expand their designs to office furniture, indoor home furniture, and furniture for
institutions such as schools and airports.

By 1957, the year in which Albinson was beginning the work on the Aluminum
Group, aluminum had become much more available and much cheaper. Global
production had grown tenfold since the Great Depression with prices falling sufficiently to make the metal "only" (in the words of Marilyn Neuhart) "three times
as expensive as finished steel and one-third the price of copper. By 1968 it was
down to 26 cents per pound, compared to its $11 per pound cost a hundred years
earlier. Once a precious metal too expensive to be used for everyday items, it
[had] become more economical to use than steel or cast iron."[30]

Design Complexities: Integrating Aluminum

Although the furniture line is known as the Aluminum Group, aluminum is
not the only material employed in its construction. The other materials have
varied over time. When Herman Miller sold the first aluminum lounge chairs in
1958, they were available with striped blue, gray, green, or brown saran weaves
(a plastic cloth). Later, the Eames Office substituted a heat-sealed Koroseal
(Naugahyde) for the saran. Finally, Marilyn Neuhart wrote in 1989, "an upholstery 'sandwich'—a front and back layer of Naugahyde or fabric, an inner layer of
stiff vinyl-coated nylon (Fiberthin), and a ¼-inch layer of vinyl foam—ultrasoni-
cally welded together in parallel transverse ribs at 1 7/8-inch intervals (the melting vinyl of the foam and Fiberthin binds the laminated assembly together) was
adopted as the production standard for the chair."[31]

Due to extensive interest in the Aluminum Group, employees of the Eames

Office are on record about how the furniture was designed and produced. In addition to interviews with Charles and Ray Eames, Marilyn Neuhart interviewed Don Albinson, providing an unusually detailed historical record of the furniture's provenance. "Charles," said Albinson, "was his normal, obtuse self, and after a short, meandering monologue about side members, and holding the seat and back in tension between them, he left it up to me to work it out." Neuhart identified Albinson's "great value" to the Eames Office as being that he "was willing to try anything," including attempting to use aluminum casting for the new chairs.[32]

The Aluminum Group required, as does all Eames furniture, a substantial investment in hand finishing and assembly, and the development of a number of specialized tools, which Albinson designed for Herman Miller in his own shop after leaving the Eames Office. He devised several "fixtures" to attach the seat membrane to the side members and the antlers to the castings. Although the assembly steps and the tools to perform them have been modified, adapted, and simplified over the years, the basic process remained the same after production shifted from the Eames Office's Venice, California, location to the Herman Miller facilities after the Eames Office closed.[33] These production runs were a step above handmade, but well below the mass production of competitors such as Steelcase.[34]

Dale Bauer and Bob Staples of the Eames Office assisted Albinson. Albinson remembered that the process was "quick, easy, and went through nice and smooth," but "it wasn't because it was a snap. We were pushing several new frontiers. We were going to stretch the cover, and hold the stretch between two funny-shaped castings—long side members—something we had never tried before."[35]

Neuhart noted that the Eames Office had prior experience with aluminum (including her husband, John Neuhart's, design of the Alcoa Solar Machine in 1957) and had used cast aluminum sections for a pedestal base for the chair shells in the low-cost furniture competition and for the lounge chair and ottoman bases.[36] After determining that it would use sand casting to make the prototypes for the structural elements, the Eames Office worked to develop the shape of the entire chair, including how its component parts fit that shape, and then experimented with the system by which the sling seat and back could be held securely in tension.[37]

The Eames Office used a variety of techniques to shape aluminum. Albinson used sand casting, which uses molds of special sand mixtures, to produce intricate shapes, large shapes, and small-quantity runs. Plaster mold casting is a refinement of sand casting that uses molds made of plaster. Permanent mold casting is a more expensive process requiring metal molds that can be used for producing thousands of castings. Permanent die casting requires the injection of molten aluminum into a steel mold by hydraulic pressure. It is used for long-run production of high-quality castings with good finishes and dimensional accuracy. Herman Miller, according to Albinson, would often use a permanent mold until the company was satisfied that it had a saleable product and then go to die cast molding, because it was a cheaper and more efficient production process that produced little waste; in the long run, it was a more economical way to go for extended production runs. Because of the experimental nature of the initial prototype building, the least expensive sand casting was the obvious choice for preproduction development.[38]

The first prototype of the modified L-shaped side members was cut out of plywood; a piece of plastic saran (a fabric originally developed for auto upholstery) was stapled to each of the two sides and held apart at the desired distance by another piece of plywood. From the plywood pattern, the next step was to streamline and simplify the form of the side members by reducing their size and depth and to add a groove that stretched along the outside of the entire curve, from the top of the piece to the bottom, into which the fabric for the seat and back could be inserted and held in tension. Through this trial-and-error process, the first casting pattern was determined.[39]

Two spreaders, which served to separate the two side members and provide the necessary structural support and stability, were positioned under the seat and at the upper back. These two curved components, which were also to be fabricated of cast aluminum, came to be known as antlers and were made in a curving U shape with the form flaring out at each end, where they were inserted into slots in the side members. Wood patterns were made of the antler sections and readied for the casting process. Bob Staples recalled making several permutations of the antlers before a shape was deemed satisfactory and passed inspection.[40]

The Eames Office contracted out the work of fashioning the aluminum. Don Albinson found a local craftsman named Mario—whose shop, Mario's Metals, was located in downtown Los Angeles—through a listing in the Yellow Pages. While Albinson could not remember Mario's last name, he did recall him as "a Mexican emigrant" who had learned the craft of aluminum sand casting in the United States and as "a true hero" to the office, which frequently used his services. "Mario," said Albinson, "was the kind of sand caster that I was a model builder. I loved him and couldn't do enough for him."[41]

"What we did," Albinson recalled, "was to take the wooden pattern for the side members and split it down the middle. Half of it was glued on one side of a board and the other half on the other side. Mario then put the board into a flask, packed it full of sand, turned it over and packed the other side full of sand. He then opened the flask, took out the board with the pattern on it, closed the flask again and poured molten aluminum by hand into the cavity. Whap! That's a sand casting." When asked how the whole assemblage stayed together and how it was then released from the sand mold, Albinson recalled that Mario "first put talcum powder on the wooden pattern, then put in a little damp sand along with one-inch pieces of window screen so that when he packed it, he had the screen to help in pulling the rib out. It's called 'green casting.' To make these things so that I could use them was a hell of a work of art. Mario made several sets of each piece, and without him, I don't know how we would have done it. It would have cost a fortune to machine the parts, and take forever."[42]

After the metal in the mold cooled, the flask was opened and the aluminum part removed. The flash line and the roughness of the metal had to be smoothed out by hand on a metal polisher—a dirty and time-consuming job. A new mold had to be made for each piece to be cast. The antler sections were cast in plaster molds and then refined and polished also. The surface of the metal was otherwise not treated.

Albinson did not recall exactly how the work on the seat progressed. It was, he said, "probably a give-and-take process between all of us." At the outset of the project, Charles had been vague about what he envisioned for the seat, and Albinson said it "just gradually developed" over time into a "membrane" stretched between the two side members, rolled up and over the top and bottom of the two curved side pieces, and fixed in place with a screw and washer on each side. It was a "neat idea," he remembered, "to use saran, a tough, new synthetic material that had a degree of transparency, which was designed for the chair by the office in conjunction with Girard because it appeared that it could function well outdoors. So, we used the saran and tripled up on it in high tension areas—the back, the seat, and the head—to provide additional support and cut down on sagging. It really worked good."[43]

In an early 1958 interview for *Interiors* magazine, Charles claimed that he had sketched the "gimmick" for the seat-membrane system on the back of an envelope while on a plane trip and had returned to the office with the drawing in hand. After he offered to let *Interiors* use the drawing for its April 1958 issue, he called Albinson (who had never seen the sketch nor heard of it) and told him to supply the magazine's editor, Olga Gueft, with a copy. Albinson hurriedly drew up a schematic and sent it to the magazine, which assumed it was an original Charles Eames drawing. According to Eames, this full-sized "doodle" of a cross-section formed the "whole framework of speculation about the chair." Eames also had concerns about the use of aluminum casting to provide the chair's side members and was nervous about the inherent license in the use of a plastic material to create a form that is not "relevant to the need" and that had more important "artistic" than functional properties:

> When you've committed yourself to casting, you've committed yourself to a plastic material and the kind of freedom that can really give you the willies. If you're dealing with extrusions or rolled sections, you are really given a limitation which is pretty nice to fall back on. But in casting there are times when the definition of the problem is pretty vague. At that moment, you find yourself face to face with sculpture, and it can scare the pants off you. There's a suspicion that maybe you're doing sculpture for which there is a valid, practical need—a need you've neglected in the past somewhere along the line.[44]

Eames's concerns included ones a scrap dealer would appreciate. "But perhaps the real question that you must ultimately face is: Is it a function of the necessary connections? In architecture, or furniture, or jackstraws, it is the connection that can do you in. Where two materials come together, brother, watch out!"[45]

Those concerns stated, the Eames Office proceeded with the mixed-material furniture using cast aluminum. When Albinson was questioned about how well the saran-seated chair functioned in an outdoor environment, he recalled that "we never got to the point of really testing its vulnerability because the chair was so expensive that nobody wanted it in saran, and they wanted an upholstered or leather or vinyl seat. It [the saran] also looked a little flimsy." There were later some reports and complaints that the saran became fuzzy and rough after prolonged outdoor exposure. The saran-seated chairs, which were made in striped blue, gray, green, and brown weaves, were sold for a time and then discontinued. The indoor-outdoor chair, like its wooden and fiberglass counterparts, became a strictly indoor item.[46]

Since the saran proved unsuitable for the durability of the furniture, Albinson switched to a heat-sealing technique to produce a thicker, quilted membrane made of Naugahyde, a vinyl "fabric," bonded to a thin layer of knitted cotton jersey. Naugahyde has a leather-like look and is often used in upholstery in place of leather. The product had just been introduced by the US Rubber Corporation. Adapting the heat-sealing technique to this material was another first for the office. "When we started the aluminum chair, we didn't have any idea about heat sealing," recalled Albinson. "It was at the beginning of the technology." And indeed, after Albinson, Bauer, and Staples had worked out the basic technique, they made the first production tooling for the bonding procedure.[47]

The shape of the seat was cut from four layers of different materials—the front and back outer layers of Naugahyde, a sheet of vinyl foam, and three strips of Fiberthin. The strips were placed in the pad running from side to side to give support to the small of the back, the top of the back, and the front of the seat. The sealing procedure on the production line was carried out in a big hydraulic press with steel platens. "It was like a big Kazaam [magic spell]," Albinson said.

> You inserted the fabric sandwich, closed the press, and clamped the dies under pressure while the heat sealing took place. The heat cycle was about 20 seconds. The layers were bonded together and then allowed to set up to make sure that it was solid. It was a terrific production technique; it had a ½-inch-wide solid strip that sealed it, and we stitched the stiffening fabric along the two sides and trimmed off the rest. Once you have the pad, everything was determined—all of the stitching and trimming patterns—we didn't have to draw lines or measure or anything.[48]

The final version of the bonded sling had a pattern of linear indentations produced by the heat and pressure, which were placed horizontally 1 7/8 inches apart down its entire surface. In addition to bonding the layers together, the heat-sealed indentations helped to define the shape and give it visual interest. After determining its feasibility, the technique was adapted to fit the various sizes of the seat pads used for the different versions of the chair. At the outset of production, the pads were heat sealed by an outside source and delivered to Herman Miller, where workers completed the sewing and assembly.

Albinson described the assembling of aluminum into completed furniture as a process requiring significant hand labor. Once the polished, cast side members (with the end holes already threaded into them) were delivered to the Miller factory from the aluminum caster, additional holes were drilled by hand on their inside edges, to serve as tacking points and placement guides for the back and the base spreader bars or antlers. Holes were also drilled and countersunk to receive the flathead machine screws used to connect the optional arms. Each size of chair (and the ottoman) required its own drill fixture.[49]

After the completed heat-sealed pad was set, front side down, on a framework on which the two side members were mounted, along with a third member that served as a center support, the pad's correct position was located, and the fiber strip on each side of the pad was pushed into the slot running the length of the casting. After both sides of the pad were secured to the castings, the assembly was "turned inside out" by rotating the side members toward each other. The seat assembly then moved to another fixture designed to hold the side members securely in position, while the corners of the pad were wrapped tightly around at the top and bottom of the castings, and held in place by a toggle clamp fitted out with a band of rubberized fabric held by a U-shaped piece of spring steel. The insertion of the Naugahyde required softening the material with an infrared heat lamp, while the toggle clamp held the end of the side member and the pad firmly in place. The long edge of the pad was then pushed into the hole at the end of the casting by the assembler, using his thumb to gradually ease the pad down. The edge was held in place by a large, specially designed aluminum washer and a stainless steel flathead Allen screw inserted into the threaded hole in the casting. The connection was then tightened so that the pad corner would not slip out. All four edges of the pad were treated in the same way. (This detail was one of the innovations in this line of furniture.)[50]

After completing the pad and side member assembly, the base and antlers were inserted into the inside grooves of the two side castings using the predrilled pilot holes. A spreading tool—an automotive jack fitted with special brackets welded at each end to fit into the grooves of the casting—forced the two side castings apart. One end of the antler assembly, which included the previously attached pedestal base, was then inserted into one of the grooves of the casting. The spreading tool was used to stretch the pad until the antler extended out slightly past the opposite side member. A hardwood stick was then used as a lever to force the second end of the antler into place. Using a rawhide mallet, the assembler tapped the antler casting into place, aligning it with the marks made at the beginning of the process. The screws designed to hold the optional arms were inserted before the antler-base assembly was attached. After the whole procedure was repeated to insert the back brace, the spreader was removed. Using the pilot holes as guides, the screw positions where the sides and antlers were attached were drilled for a tap to receive an Allen set screw. The chair (without arms) was then complete. Tilting and pivoting mechanisms were also added to make the chairs commercially viable for office and institutional customers. A small head cushion made of vinyl was affixed to the back of the pad at the top of the larger lounge chair in the series and folded over to the front side. The back stretcher could also be used to move or carry the chair.[51]

After experimenting with adding an armrest, a cast aluminum member in the shape of a modified rectangle with rounded corners became the production model. The armrest followed the curve of the side member as it turned around the seat section and projected out at its top leading edge in a modified, rounded point. It was attached to the side member by three flathead Phillips machine screws set through the holes drilled into them and then into the three extensions cast with the armrest. Special molded, black-dyed nylon sleeves were inserted between the side and the armrest to position the arm away from the chair.[52]

The same process was used to assemble the accompanying table, which employed one antler. This piece also used the quilted membrane of radio-frequency-sealed (RF-sealed) Naugahyde stretched between two side members and held taut by the same method as the chairs. The matte-black steel pedestal and the polished aluminum four-pronged bases were adapted from the bases used for the Eames lounge chair and ottoman. The chair pedestal was inserted into an extension in the cast-seat stretcher. Later, in the mid-1960s, the Eames universal base was substituted and became the production standard for both the chairs and the tables.[53]

The RF-sealed black Naugahyde pad was the production model. When customers began asking for pads in Herman Miller textiles, instead of Naugahyde, the factory attempted to use the radio frequency technique to bond them, but discovered that any foreign matter embedded in the textile burned and caused holes to appear—making it an impractical solution. The office contacted Albinson (who had left Eames in 1959) to see if he could find a way to bond fabric pads, and Albinson agreed to do so in his own shop. Albinson's description of the means by which he arrived at a working system is a good example of his pragmatic, commonsense approach to adapting existing tools and fixtures to new uses:

> I finally figured out a way to do up small sample pads with a method that utilized two square electric frying griddles that contained a built-in thermostat, a flat surface, and heating elements already molded into their bottoms. I cut the sides off the griddle top and bottom, screwed strips of aluminum angle onto one of them and laid the fabric sandwich between them. After the sandwich was laid up we would clamp it between the two heated plates, one flat and smooth and one with the ribs on it. The heat was distributed evenly and after fooling around with the times and temperature, we found out we could heat seal in about eight or 10 minutes. Using the sample of heat-sealed fabric as a guide, we then proceeded to make a tool to make the full size pads. . . . I

figured we could form these pads by just using ¼-inch aluminum plates, top and bottom, instead of buying a big $25,000.00 press. We put our fabric in, closed it, and sealed the outside edge with a rubber flapper of silicone rubber, which is not affected by heat. Once you close it, and the little flapper is sealed, we turned on the 40-dollar vacuum pump, drew the air out of it, and we have 20 tons of pressure—perfectly spread out. You have atmospheric pressure pressing the same over the whole surface, so you just make different size frames for different size pads. The aluminum plates have strip heaters screwed to them and a thermostat. It sounds complicated, but it isn't. I made the press and the frames for all the different sizes of pads in my shop. And we did it all for about $2,000.00![54]

The chairs in the Aluminum Group of furniture included an upright chair with a medium-high back that could be used as a desk or a dining chair (with or without armrests), a lounge or easy chair with a high back (with or without armrests), and a second lounge chair with a higher back, armrests, and a head cushion. The cast side members of each chair were formed in a different angle. The aluminum was either brightly polished or coated with a dark coating. The chairs' sling seat was available in eight fabrics and vinyl. The cast aluminum pedestal-support bases, which were originally trapezoidal in section, were later changed to an elliptical section.[55]

The line also included two round tables, one at coffee-table height and the other at dining or working height. Tabletops were available in marble, slate, and white glass and were supported by the same pedestal bases used for the chairs. Later variations included hardwood veneers and plastic laminates. Both Herman Miller and Vitra still offer these tables in a wide range of finishes and options.[56]

In 1989, the Aluminum Group consisted of a high-back, tilt-swivel lounge chair (with or without arms), a low- or high-back, tilt-swivel desk chair with an adjustable seat (with or without arms), an ottoman, a coffee table, and a dining table. Neuhart, Neuhart, and Eames noted, "Each of the chair's side ribs is a curved, one-piece, die-cast aluminum member; two 'flaring spreaders,' or 'antlers,' are screwed into the frames at the back and under the seat, connecting the two ribs."[57]

In this way, the aluminum was integrated with the synthetic materials used in the upholstery. This was done, Neuhart, Neuhart, and Eames noted, by "sewing a stiff strip of plastic (Royalite) along the edges of the upholstery and then working it into the grooves with the frames turned inward. The frames were then flipped over, pulling the fabric in place under tension. The aluminum spreaders [were] screwed into position to keep the chair sides a fixed distance apart. The back stretcher [was] also a carrying handle and the seat stretcher a support for the pedestal base."[58]

Institutional Applications: Tandem Sling Seating

The expensive Eames aluminum designs became coveted in offices; they also found their way into institutional settings where millions of people throughout the industrialized world who could not or would not choose to spend the money on individual pieces have used them. In addition to developing furniture that would withstand the elements on a patio or that would express luxury in an office, the Eames Office used its techniques to put aluminum furniture in institutional contexts. It did not create thousands of cheap chairs as it intended with its first attempts at aluminum chairs (which failed). Instead, the mixed-material designs of the Aluminum Group were transformed into shapes suitable for institutional use. Bob Staples noted that problems articulated by a client, such as an airport or school, sparked innovative solutions by Charles Eames. Staples remembered: "I think [Charles's] best product either came from his own head or from some legitimate client's request. Not just from Herman Miller, but when [the architects] wanted a new chair for the Chicago airport." That was what Staples called "a love investigation"—a passionate exploration of the idea.[59]

The Eames Office began to experiment with mass-seating systems for schools and other institutions in 1954. The first attempts followed the designs of stadium seating, using a single steel support to join a series of seat shells. In 1958, the architect Eero Saarinen asked the Eames Office to address the need for comfortable, sturdy, and attractive public seating for Dulles Airport in Washington, DC. At the same time, C. F. Murphy Associates, the architects of two new terminals at Chicago's O'Hare Airport, were searching for good-looking seating that was strong, durable, resistant to wear, and easy to maintain and repair.[60] C. F. Murphy Associates contacted Robert Blaich, the director of Herman Miller's Special Products Division. Blaich asked the Eames Office to address the problem by making use of its experience with aluminum casting.[61]

Eames Office members Dale Bauer, Peter Pearce, Robert Staples, and Richard Donges worked on the airport seating system with assistance from Gene Takeshita. They used the same design methods as those used by the Aluminum Group. Neuhart and Neuhart described the seating system as consisting of single or double rows of two to six seats, and double rows of 10 to 12 seats secured to a continuous steel T-beam. As in the Aluminum Group of chairs, the side members are made of polished cast aluminum, as are its pedestal base. The interchangeable seat-and-back sling system made of a sandwich of foam, Fiberthin, and Naugahyde also [uses] and extends Aluminum Group technology. Instead of the parallel lines of the aluminum chairs, however, the materials of the sling backs of Tandem Seating are welded together in a pattern of trapezoids and triangles designed by Bauer. The slings are tension-mounted between the side and seat members, which are assembled with mechanical joints. A single screw holds the hard urethane armrest in place and affords easy maintenance for replacing the components. Although Charles felt that black was the best choice for the slings, they could be ordered in a variety of colors. A table attachment can be substituted for one of the seats.[62]

The airport seating system remains in use and is critically acclaimed. Marilyn Neuhart described its ubiquity: "It is hard to imagine that there is any air traveler anywhere who has not sat in it numerous times."[63]

The system went into production in 1962. Eames tandem seating was installed first in O'Hare and Dulles terminals and later in airports around the world. A humorous photograph of a delivery shows Eames Office member Jim Sommers sitting atop a row of tandem sling airport seats that are propped up in a decrepit pickup truck. The image demonstrates the aluminum row's strength and light weight, properties that airports valued as part of a durable solution to seating waiting passengers. Herman Miller still produces the seating units in single rows of 2–6 or in double rows of 10 and 12. The company also still manufactures a table attachment that can be substituted for a seat.[64]

A variation on the Eames solution for public seating, tandem shell seating, combines the shells developed for the plastic armchairs and side chairs with the base developed for the Eames tandem sling seating. The shells are attached to cast aluminum spiders by rubber shock mounts and then are mounted onto a black, epoxy-painted steel T beam.[65]

The Eames Office created a similar variation in 1964 for school lecture halls. "In the School Seating, the aluminum leg support base forms a triangle and, like the steel beam, is coated with black epoxy. Each chair is provided with a right-hand fold-up tablet arm laminated in a neutral light formica."[66] Herman Miller still produces this design also.

In a 2000 essay, critic Craig Vogel praised tandem sling seating as comfort-

Eames tandem sling seating being delivered shows off aluminum's strength and lightness. Courtesy and copyright © Eames Office, LLC (www.eamesoffice.com)

able and aesthetically elegant. Although more folding chairs had been made than tandem sling seating units, Vogel argued, "more people may have actually sat in the Eames chair. All of the problems associated with the tubular folding chair are elegantly resolved in this design. Using primarily the same materials—an aluminum frame and plastic seating—the Eameses created an elegant, subtle solution that fit the context of a modern airport."[67]

Vogel wrote that the designers, architects, and manufacturer of tandem sling seating "all wanted a piece of furniture that would be the quintessential expression of modern design and complement the design of both the airport terminal and that most modern form of transportation, the jet airliner."[68] For Vogel, aluminum and Naugahyde exploited their ersatz qualities in ways that exuded class and luxury. "Although it looks like it was made of stainless steel and leather, the system is actually produced at a fraction of the cost, weight, and maintenance of those materials."[69] Vogel noted that the chair system is easy to manufacture, easy to ship, easy to assemble, and easy to maintain. "The seating surface is stain-resistant and rugged and can be cleaned with a cloth, and the continuous form of the aluminum frame does not have small crevices where dirt can collect. The seating is easy to move and clean under. If the seat is damaged, one panel can be easily replaced."[70] In all, Vogel concluded, tandem sling seating "is one of the few Eames designs that [is] widely accessible to the public."[71]

Covetable Aluminum

Vogel's point on accessibility reflects not the Eames Office's ability to manufacture Aluminum Group pieces (Herman Miller still is able to produce pieces on demand), but rather the affordability of those pieces. The Aluminum Group furniture quickly became classics in the Eames Office repertoire, and they were "covetables," to use Charles Eames's term. Alcoa noted their popularity; in 1959's *Design Forecast 1*, Joseph Petrocik observed that Eames had "long predicted that aluminum would one day rank with wood as the furniture manufacturer's most valuable material."[72]

Petrocik used Eames designs to note that "brilliant dimensions of function and beauty are appearing in aluminum furniture design. They are the result of imaginative styling, technological research, the use of new casting techniques, the introduction of color through anodizing processes, the development of paints and lacquers, and the cooperation of the aluminum industry itself with the designer and manufacturer."[73]

Over the years, Eames furniture has remained popular among consumers who could afford the high prices. The Eames Office's experience with transitioning furniture from patios to offices is instructive in the economics of individual pieces and highlights the clientele Charles Eames intended to serve. In 1959, Aluminum Group chairs and ottomans with the early saran fabric slings were formally introduced in new settings in the Herman Miller showroom in Los Angeles.[74] These were immediate critical and commercial successes, spurring further creations employing aluminum. The following year, Time Inc.'s chair, Henry Luce, asked the Eames Office to design three lobbies for the brand-new Time & Life Building in midtown Manhattan. The main lobby was a rectangular space divided by a reception desk in the center. On one side of the desk was a lounge area with cast aluminum and black leather chairs and turned walnut stool-tables.[75] Ray Eames recalled that the design "combines the technologies used in other Eames furniture—molded plywood, aluminum casting, and padded leather and foam rubber cushions."[76]

Charles Eames remembered the decision to move from home and lounge furniture to office furniture:

> We were doing some chairs, which we felt you would use in a home, and, for
> example, even a swivel or tilt chair seemed to be a very natural thing to have
> in a living room as you shift your weight you're comfortable in the evening.
> And so what happened was Herman Miller had, I can remember it well, in
> Chicago where we set up the sample rooms, which we viewed as great rooms
> for all children, people, and people in business came in and looked. "Ah, this
> is just the thing we need for the office, this lecture hall, this so and so." Essen-
> tially, the functional aspects, function having to do with certain proficiency
> . . . and I suppose, assuming some of the fact that it didn't, it seemed to result
> in kind of not impossible environments. This changed the whole direction of
> the business because now there is the instinct to try to, not that there hadn't
> been the core of that in terms of, you see, the Rohde Desks, and what not,
> were in effect, some of the most successful of Herman Miller's things even at
> that time. Then George [Nelson] went on—I'd have to look at the map to see
> what time and where, but George's, those executive situations, these things
> fell right into that, so that the whole mass, the whole trend shifted, and now
> we have some things which were done as part of the office program. There's
> a whole corner of it which we feel would shift and make marvelous domestic
> things. They never made the decision of leaving the domestic things and en-
> tering the office.[77]

The office chair designs were similar to those of the original lounge chair but with the plywood core hidden within a real leather upholstered seat and back. This model is still sold by Herman Miller as part of a line called Eames Executive Seating, with individual pieces retailing for more than $2,000.[78] Production of the furniture remained in the Eames Office even as popularity increased. Charles Eames remembered: "The first 5[,000] or 10,000 pieces we actually made right here in the building [Eames Office]."[79] Regardless of the scale of production, the pieces were expensive, to Nelson's chagrin. "We didn't have any control [over prices]. We tried to do things reasonably, but Herman Miller prices, for a lot of reasons, were always shocking, no matter what we did."[80]

Reflecting on the class aspect of his products in 1974, Charles responded that the designer served the desires of the consumer. "There was a moment there before we sort of realized what was happening, when we found ourselves trying to talk young couples out of lounge chairs, which, God knows, they shouldn't have been doing, and try to talk them into sensible uses of orange crates and things we thought were so reasonable. But, so what? They wanted a lounge chair."[81]

The Eames executive chair and subsequent takes on the Aluminum Group furniture were all expensive, and most, like the original Aluminum Group, remain in production at Herman Miller. There was at least one attempt to make an Eames design more affordable. Introduced in 1964, the 3473 sofa (named after its Herman Miller catalog number) was supposed to be affordable, but George Nelson noted that the attempted production methods to make the piece less expensive did not succeed. "One of the things we did that was supposed to make things very cheap, and didn't, was the idea that if you used the new epoxy glue processes, you could then make the sofa out of small pieces, which could then be plated more economically than big ones. And that turned out not to be the case. This thing cost as much as a Volkswagen by the time it hit, and it doesn't even have a motor on it."[82] Herman Miller discontinued the sofa in 1973.[83]

In the last years of Charles Eames's life, the Eames Office focused on refining the designs of existing chairs rather than introducing dramatically original pieces. The intermediate desk chair, a small mobile chair introduced in 1968, was based on ideas used in the Time-Life chair and the executive desk chair. Like them, the intermediate desk chair used leather, Naugahyde, and cast aluminum. Like the sofa, manufacturing the chairs proved to be too expensive, and they too were discontinued in 1973.[84]

Other later designs that survived the 1973 downturn and continue to be produced in the twenty-first century include the Eames Soft Pad Group of chairs (introduced in 1969),[85] a drafting chair (1970) using an aluminum spider structure,[86] the loose cushion armchair (1971), and, finally, a sofa with arms.[87] A model for this piece was completed in 1967 and then set aside for several years. "Work on a wood and leather three-seat sofa was begun in 1976 under Charles's direction," but was not completed until after his death in 1978. In this sofa, "two identical die-cast aluminum members serve as integral base/back brace units. The arm castings are also symmetrical and have padded rests. All of the aluminum parts are brightly polished." Production of this piece began in 1984 at Herman Miller's Italian factory, and Herman Miller still manufactures the Eames teak and leather sofa.[88]

The Environmental Herman Miller

For Charles and Ray Eames, the ecological benefits of recycling aluminum were less significant than its materiality. A durable, malleable, yet light metal allowed the construction of minimal frames that could be easily mass-produced.

Recycled aluminum made the designs somewhat more affordable, allowing Eames furniture to find homes on patios, in living rooms, in offices, and in airport and corporate lounges.[89] The metal's aesthetic appeal translated not only into structural elements of furniture design, but also into finishes for kitchen appliances, such as refrigerators and washing machines.

The Eames Office's focus was on creating covetable goods that would endure and provide enjoyment. In a 1974 interview, Charles Eames articulated the goals of his designs:

> The whole point [of the idea of covetables] hinged around the fact that . . . our society, and the evidence certainly points to it, has arrived at a point of universal expectancy where everybody feels he has a right to what any other person has. Then, if that realization of these expectancies, these goods, these covetables, if it is going to be less than a real tragedy, things better change because the things we have as goods are not that good to make universal. And then we looked at the rules of what a new covetable would be. It had to do with, if it's going to be universal, then you have to be able to have a lot of it, and it doesn't deteriorate. If it's going to be a covetable, it has to be painful to get and not easy to get, or what's to covet? And if you have it, it's going to give pleasure to you, and yet, when you end up, it is that the new covetable[s] are really sort of concepts, mastery of processes or systems, or ideas.[90]

The Aluminum Group furniture was part of a larger context of acclaimed design for mass production, and the furniture that the Eameses designed for Herman Miller after 1958 influenced a wave of furniture and appliance design using the metal. Other designs employing aluminum during the late 1950s included the shelving of Alex Girard, the aluminum table of Isamu Noguchi, the sunlight lounge chair of British designer Julian Herbert, and the dining room chairs of Peter Moro. Encouraged by incentives from industry, designers on both sides of the Atlantic incorporated aluminum into their work. Crucially, this adaptation resonated with consumers. Herman Miller furniture was expensive; instead of being perceived as flimsy and fake, Eames Aluminum Group designs were embraced as modern and elegant.

While continuing to produce many of the most recognizable designs created at mid-century, Herman Miller now emphasizes sustainability in its production methods and operations. In this way, the company offers perspective on how upcycling secondary materials has links to both the past and innovative approaches.[91]

But Herman Miller and its designers are not alone in this project. In the half century since the company began selling Aluminum Group furniture, aluminum has become a structural material for furniture from designers all over the world, including Philippe Starck of Paris, Norman Foster of London (whose 20-06 chair for Emeco, he emphasizes, is made of 80 percent recycled aluminum), Joris Laarman of Amsterdam, and Charles Pollock of Brooklyn.[92] Designers in the twenty-first century use aluminum for both stylistic and environmental reasons. Emeco head Gregg Buchbinder quoted Starck's philosophy as "heritage begets recycling," explaining that his company's goal is "to take things out of the landfills and make high quality things."[93]

Herman Miller has adopted the language and production practices of sustainability. In 1986, Hugh De Pree argued, "We have concern for the environment in which we work. The property and facilities we develop for our use must improve the quality of life in the communities we serve."[94] He quoted Herman Miller's model maker Pep Nagelkirk, who asked the key questions about any Herman Miller product: "Have we built into it durability? How will it look in ten years? Does it have value?"[95]

Longtime design director George Nelson was especially vocal about the environment in the last 20 years of his life; he focused on environmental worries in a 1974 oral history, and a 1987 memorial booklet featured him expounding on the subject at length. Nelson argued that functionality included environmental health, and the process of design at Herman Miller stressed this approach. "There's a moment in the evolution of a product when everybody is very concerned about the way it looks, there's no doubt about it. If the preoccupation with looks gets out of hand, then somebody holds up his hand and says, 'Wait a minute, the thing, really, ought to work.'"[96]

"To design a less bloody future," Nelson said, "we have to learn how to become good gardeners. A planet tended by a race of gardeners would remain in very good condition for a very long time. We also need more people who know how to design. Creative people. Every generation produces only a handful, so maybe we will have to learn to grow them.[97]

Nelson worried about nuclear annihilation and environmental destruction, noting that designers needed to better cope with "what we have, which is a rather beat-up planet which still has a lot of life left in it. What this means in essence is we better learn how to act as gardeners instead of miners. This is one way of saying it."[98] "See, if overnight we woke up and everybody was a gardener, in a general sense, the possibility that we can live on indefinitely on this planet is extremely good. It would mean getting rid of a lot of things that we've considered essential, but we can do it."[99]

Nelson hoped that young people would choose this more responsible practice and say: "Gee, I better put my bets on number three and see what I can do about becoming a planetary gardener. And there are a lot of people working at this in one way or another. I'm not talking about agriculture, obviously."[100]

Nelson's contrasting gardening to mining speaks directly to the aversion to extracting primary metals. Speaking in the mid-1980s, a time when recycling gained popular acceptance as a more environmentally responsible way of sourcing industrial materials, his framework reflected Herman Miller's embrace of sustainable design strategies.[101]

In the years after De Pree's and Nelson's time at Herman Miller ended, the company built on their environmental concerns. Nelson's influence helped Herman Miller establish a comprehensive environmental program in the 1980s, practicing extended producer responsibility (EPR) through its AsNew program (which combines recovered components with new components) as early as 1984. Herman Miller began working with William McDonough in the 1990s to establish a Design for the Environment (DfE) program and develop cradle-to-cradle manufacturing. This included the design of the Greenhouse manufacturing plant and office space.[102]

By 2002, Herman Miller's efforts in establishing EPR merited inclusion in *A Handbook of Industrial Ecology* edited by the pioneering industrial ecology scholars Robert U. Ayres and Leslie W. Ayres. In a chapter on extended producer responsibility, John Gertsakis, Nicola Morelli, and Chris Ryan praised the company for developing comprehensive environmental programs aimed at eliminating or minimizing life-cycle environmental impacts, particularly those resulting in solid and hazardous wastes. "Design for disassembly principles have been embodied in their furniture to help ensure that service and repair, end-of-life remanufacture and materials recycling are not only realized but add value to the overall enterprise."[103]

In 2006, Herman Miller's DfE program manager, Scott Charon, and DfE manager, Gabe Wing, collaborated with McDonough Braungart Design Chemistry's senior project manager, James Ewell, and Clean Production Action's research director, Mark Rossi, on a *Journal of Industrial Ecology* article detailing Herman Miller's efforts to meet cradle-to-cradle design goals.[104] "Working with MBDC," Rossi and his colleagues wrote, "Herman Miller developed the DfE product assessment tool, which evaluates the extent to which a producer meets the goal of the cradle-to-cradle ideal—that is, made from 100 percent biological and/or technical nutrients."[105]

This evaluation method has altered the composition of Herman Miller designs, including eliminating PVC components and increasing recycled content (with the goal for all Herman Miller products to attain a recyclability rating of 75 percent); the products are also designed for rapid disassembly using common tools. Metals play a crucial role in these efforts. Rossi and his colleagues concluded: "Herman Miller is working with its suppliers to maximize recycled content in its steel and aluminum products."[106]

Herman Miller also participates in Cradle to Cradle certification of its products, complies with Global Reporting Initiative efforts on corporate transparency, and releases an annual "better world report" in which it assesses its effects on society and the environment. The company has a goal of zero hazardous waste emissions in its manufacturing processes by 2020 and expects to reach that goal while continuing production of the furniture Charles and Ray Eames pioneered in the 1950s.[107]

In the three decades since Nelson's death, Herman Miller has continued its quest for sustainability, not only sourcing secondary materials, but also hiring William McDonough + Partners to develop its production facility. Herman Miller claims that its Greenhouse produces just 15 pounds of landfilled waste each month, even though it produces a made-to-order chair every 13–21 seconds.[108]

"Highly Sustainable Design"

Herman Miller's approach is not unique in industrial production. Samantha MacBride noted that voluntary efforts at a variant of extended producer responsibility known as "product responsibility" gained favor in many industries in the 1990s.[109] In industrial furniture design, Herman Miller's efforts have been joined by similar approaches by Emeco and high-profile designers. In the twenty-first century, pieces by Norman Foster and Philippe Starck explicitly use aluminum as an integral part of a sustainable design approach. Both men were influenced by the work Herman Miller developed at mid-century. The British architect and designer Lord Norman Foster is a disciple of the Eames Office, which he encountered while in the United States on a postgraduate fellowship at Yale in 1961. In addition to designing skyscrapers, airports, and other major buildings across the world, Foster + Partners also produces furniture.[110]

Foster was attracted to the possibilities of aluminum early in his career, evidenced when he identified the Boeing 747 as his favorite "building."[111] On his return to Great Britain, he applied what he had learned from the Eames Office to his own architecture and furniture design. The Design Museum noted the explicit influence of the American designers in Foster's Reliance factory, with its elegant exposed steel structure with diagonal bracing.[112]

Foster's twenty-first-century furniture designs, like those of the Eames Aluminum Group, use the metal to fashion durable, elegant chairs of considerable worth. His 20-06 stacking chairs, produced for the Electric Machine and Equipment Company (Emeco), retail for $625 apiece. His processes and end results, however, are more explicitly framed as environmentally sustainable than anything Charles Eames or his collaborators articulated in the 1950s. First sold in 2006, the chairs were designed with recyclability in mind; Emeco's promotional materials note that they consist of 80 percent recycled aluminum, with roughly half that share coming from beverage cans and the other half from prompt industrial scrap. Architectural journalist Marcus Fairs noted that Foster merged sustainability with aesthetics. "The chair has been engineered to use as little of the material as possible, although this is done more for aesthetic than ecological reasons. But the chair has an estimated lifespan of 150 years, making it a highly sustainable design."[113]

Philippe Starck rose to fame in 1982 when he was commissioned to design President François Mitterrand's private chambers. His use of aluminum may also reflect the fact that his father designed airplanes. Emeco of Hanover, Pennsylvania, sells several Philippe Starck designs made entirely of aluminum. The designer and the furniture company collaborate on both new designs and reinterpretations of older models. Emeco began producing its all-aluminum 1006 chair, featuring a side rail welded to the back legs and sophisticated and complex curves to the stretchers, for the US Navy in 1944.[114] Starck reinterpreted the chair successfully enough that a model initially built for military use found its way into the restaurant of New York City's Paramount Hotel. Impressed, Emeco asked Starck to create new models that were both "environmentally responsible to produce and sympathetic to the original 1006."[115]

The results, beginning in the year 2000, included the Kong and Hudson models. Starck found inspiration in the furniture of Versailles when creating the Kong chair for a Chinese restaurant in Paris. It sells for $1,665 per unit. A Hudson desk chair of polished aluminum sells for $2,205. Others of his all-aluminum designs

Emeco Hudson barstools. Project: Lobby lounge at Marriott Marquis, Washington, DC; design: HOK; photograph by Adrian Wilson. Courtesy Emeco

sell for hundreds or thousands of dollars apiece. In the sales description of the Hudson task chair, Starck remarked: "Working with Emeco has allowed me to use a recycled material and transform it into something that never needs to be discarded—a tireless and unbreakable chair to enjoy for a lifetime. It is a chair you never own, you just use it for a while until it is the next person's turn."[116]

Sarah Nichols observed that Starck's designs are designed for disassembly. "Philippe Starck's Louis 20 armchair consists of aluminum arms and back legs that are screwed to the polypropylene seat and back. It can be dismantled into recyclable elements in a matter of seconds simply by undoing the screws."[117]

The designs Foster and Starck create have continuities and differences with the Eames furniture Herman Miller produces. One crucial difference between the Emeco chairs and the Aluminum Group series is that the former are entirely fashioned from aluminum. Should one of Foster's or Starck's chairs break, and should the owner decide that the chair is unworthy of repair, it could be scrapped easily without the need to separate leather, fabric, or other materials. The Emeco chairs not only use scrap aluminum as a source, they can easily be returned to scrap without worries about creating waste products beyond those generated in the normal production of secondary aluminum. Actually scrapping one of the Emeco chairs would be economically unsound, however, given the value of the designed object, and it is likely that most will serve as chairs for more than a century, reflecting Fairs's estimated life-span for the object. The Emeco chairs are conscious designs that upcycle secondary material.

Is the Aluminum Group upcycled? The secondary aluminum used to fabricate the furniture has its economic value transformed due to the prestige of Eames design and Herman Miller manufacture. The result is a durable good that retains its value: 1958 chairs sold in 2014 for more than $1,500 apiece, and newly manufactured versions of the chairs also sell for more than $1,000. Should the chairs become damaged, the aluminum frames could be melted and refashioned into new furniture, including new Eames chairs.

Not all of the material in Aluminum Group furniture is freely recyclable, however. The Naugahyde and foam used to provide soft seating is embedded with toxins that might be released into the air or groundwater in the act of separating the aluminum from them. Moreover, neither material can be simply transformed into new versions without degradation. Had the EPR strategies Herman Miller put into effect at the end of the twentieth century informed Eames designs in the 1950s, these materials would not have been included in the original furniture. (Even with Herman Miller's current EPR, the public could be provided with more explicit information on the toxins generated by the company's manufacturing processes.) Yet with this caveat, the aluminum furniture Herman Miller has produced for more than half a century may be said to take secondary material and transform it into durable goods of greatly enhanced economic value.

The history of aluminum furniture reveals that environmental concerns have influenced design strategies since the 1970s. However, the resulting products retain similarities with the Aluminum Group furniture created in 1958. The Eames designs, like more recent ones, employ secondary material. Designs from both eras produce furniture that is expensive to buy new but that retains its value over time. These durable goods are worth more to reupholster or repair than they are to process as scrap; thus, these applications of secondary aluminum remove technical nutrients from industrial production. Covetable chairs use aluminum, but do not provide aluminum to feed the creation of more covetable chairs. The hunger for new Eames chairs can easily be satisfied by the click of a button on the Design Within Reach website; more people than ever have access to these beautiful designs. All one needs is a credit card and an internet connection—and Herman Miller needs enough aluminum to meet the demand for its covetable products. That this aluminum rarely comes from older Herman Miller furniture complicates the notion of the circular economy; the life cycle of covetable aluminum goods merits further examination.

Chapter Six

Guitar Sustain

H ow can three pounds of aluminum be worth $312,000? The metal has many applications, but the price of that amount of scrap aluminum on the open market is about $2. The answer has something to do with Charles Eames's notion of covetables. Three pounds of aluminum sold at auction in 2007 for $312,000 not because of the material properties of the metal, but because the metal was part of a guitar owned and played by the Grateful Dead's Jerry Garcia in the mid-1970s. The chance to possess an instrument owned by a dead rock star escalated the material's price within the designed good far higher than its market price as scrap. The sale was notable enough to be mentioned in the *New York Times*'s obituary of the man who built the guitar, the California-based designer Travis Bean.[1]

Admittedly, celebrity cachet accounted for the majority of the auctioned guitar's value, but this make of guitar is very valuable regardless of which musicians play it. Travis Bean guitars regularly sell for several thousand dollars, far cheaper than Garcia's instrument, but much more expensive than most new electric guitars. Why are these aluminum-necked instruments so expensive?

Early Aluminum Guitars

Aluminum-necked guitars reveal much about the values placed on new and old goods in ways that expand on Charles Eames's goals for his furniture designs. Aluminum has had an important role in the history of musical instruments. Luthiers recognized the durability and resonance of aluminum as early as 1928.

Vaudevillian guitarist George Beauchamp sought a louder instrument for stage work and asked the Los Angeles–based violin repairman John Dopyera for assistance. Dopyera and his brother Rudy devised a prototype using three thin, cone-shaped aluminum resonators, and the three men began producing the "resonator guitar" for the National String Instrument Company. This acoustic instrument remained in production both at National and (after a dispute) at the Dopyeras' new Dobro Corporation. In the legal battles that produced the split, Beauchamp was fired from National. He partnered in 1930 with National's factory superintendent, Harry Watson, to develop a "frying pan" (its circular body resembled a frying pan) electric lap steel guitar ("steel" referring to a metal bar that players use to change the pitch of the strings, not to the material of the instrument) made of cast aluminum.[2]

The Great Depression hindered Beauchamp and Watson from marketing their new design, but Beauchamp subsequently partnered with Adolph Rickenbacker in a new company initially named the Ro-Pat-In Corporation, which eventually became Rickenbacker (spelled "Rickenbacher" until the 1940s). They began selling the cast aluminum Electro A-22 in 1932. The success of the A-22 led to more lap steel guitars employing aluminum bodies; in 1935, the Gibson Guitar Corporation produced 115 cast aluminum E-150s (named because it sold for $150) before abandoning aluminum for wood. Rickenbacker stopped advertising the A-22 in favor of more popular instruments, but continued to manufacture the guitar through 1950 and, after discontinuing production during the Korean War, a redesigned model from 1954 to 1957.[3]

Aluminum was an unusual material for a stringed instrument; centuries of working with wood to create acoustic instruments combined with aluminum's relative expensiveness to limit its use among manufacturers. The A-22 was aluminum's greatest success; the novel instrument became a staple of Hawaiian music as well as a precedent for the lap and pedal steel guitars adopted by country-and-western musicians after the war.[4]

The abundance of affordable aluminum was one of two critical developments in the history of aluminum guitars. The other was an advance of designs for instruments that were held and played like acoustic guitars, but designed with electrification in mind. The first electric guitars adopted the hollow bodies that were standard in acoustic instruments. Some luthiers, notably the Americans Les Paul and Leo Fender, found that solid wooden bodies could change the performance of the instrument, reducing feedback and enhancing the sustain of struck notes. The popularity of these guitars (the Les Paul models manufactured by Gibson and the Telecaster and Stratocaster models built by Fender) among jazz, blues, country, and rock musicians altered the expectations and market for electric guitars.

Wandré

As solid-bodied electric guitars grew in popularity, the availability of affordable aluminum allowed experimentation with the metal beyond the lap steel models developed in the 1930s. The first major designer of electric guitars employing aluminum lived and worked in Cavriago, Italy. Antonio "Wandré" Pioli was a self-described sculptor, artist, and motorcycle enthusiast who was the son of a luthier. Pioli saw the advantages of aluminum in vehicle design and sought to replicate the durability and styling of postwar vehicles in the instruments his father made. Beginning in the mid-1950s, Pioli produced a great variety of aluminum-bodied electric guitars under the names Wandré, Avalon, Dallas, Davoli, and Noble, among others (often depending on the distributor in different countries around the world). The Wandré name was the most widespread, with models including the Krundaal Bikini guitar (featuring an amplifier and speaker built into the guitar's body) and the Wandré Doris, a solid-bodied guitar with a tremolo constructed from salvaged motorcycle parts, push-button controls for activating the guitar's pickups that resembled the radio buttons in an automobile, and styling that varied by individual instrument.[5]

The variations on the Doris reflected the eclectic experimentation Pioli applied to his instruments. Aluminum allowed him to develop body shapes ranging from minimal to ornate, and a book featuring representative designs would be thick. Although the bodies varied widely, one constant was the use of aluminum for the instrument's neck (the area joining the headstock and extending to the body of the guitar). This decision had implications for the sonic performance of the guitars; aluminum's strength and density exceeded those of the densest woods Gibson used in its Les Paul models. A "neckthrough" or "set" aluminum guitar (the aluminum neck was built into the body of the guitar and provided one continuous aluminum span from the headstock to the body) could extend the sustain of notes played on the instrument. Guitarists and bass guitarists who valued that aspect found the performance of aluminum-necked instruments especially attractive. Aluminum had one other distinct feature. Aluminum guitar necks do not warp due to humidity and temperature variation, problems that require substantial maintenance or replacement in wooden-necked guitars. Aluminum necks could feel cold or hot, and thus temporarily alter the tuning of the instrument, but they remained straight. Players able to adjust to the different tactile experience of aluminum enjoyed the instruments' reliability.[6]

Between 1956 and 1969, Pioli's guitars sold throughout the industrialized world, with pockets of popularity in the United States, Argentina, the Netherlands, and Italy. Although Pioli lived until 2004, he ceased creating new instruments in 1970, when he sold his guitar factory to establish a leather clothing business. One aficionado estimated that Pioli produced about 70,000 Wandré instruments during his career.[7]

Instruments passed through the secondhand market in the years after Wandré ceased production. But Wandré did not have the mass appeal or marketing of Fender Stratocasters, and the guitars became curiosities. Texas-based country music guitarist Buddy Miller happened across a white Wandré guitar in the window of a Boulder, Colorado, pawnshop while on tour in 1976. Advertised for $85 ($354 in 2015 dollars), the guitar was unusual enough that "I thought it would look pretty good on my wall, it had sparkles, so I offered the guy $50, he said 'Sure.' When I took it to the gig for a joke and plugged it in, it sounded real good. When I got back to Austin, I ordered the yellow pages for Boulder and went through the pawnshops till I found the place again. They had four more, so I bought them all. They used to import them. It says Noble on the top of it, but that's just the name of the accordion importer in Chicago who brought the Wandrés into the country."[8]

Surprised by the performance of these unusual guitars, Miller adopted them as the primary instruments he has used for 40 years. "They're real good guitars. I've got the most conservative ones. A few of the really weird ones have become expensive, and are valued as works of art."[9]

Early American Designs: Burke, Messenger, and Veleno

By the time Miller found his Wandré in the Boulder pawnshop, American luthiers had started to build aluminum-necked guitars. In the United States, small production runs were made by Oregon-based designer Glen Burke's Tuning Fork Guitar Company and by Orion guitars in the early 1960s. Burke filed his patent for an aluminum-necked guitar in 1960, using his design to produce mostly 6-string guitars and some 12-string guitars with bodies decorated with a wide array of details ranging from paint to cowhide. Kahoots guitarist Elisha Wiesner described Burke's guitars in 2013 as "next to impossible to find" on the vintage guitar market.[10]

In 1967, a West Coast company named Messenger began manufacturing a guitar with a magnesium-aluminum alloy neck that extended as one piece from the headstock through the guitar's hollow body, bolting on at the neck-body joint and endpin. Messenger went out of business in 1968, ending a run most famous for having a guitar eventually owned by 1970s Grand Funk Railroad guitarist Mark Farner.[11]

John Veleno had greater success working with aluminum. Veleno began playing guitar in Massachusetts in the late 1950s, later working as both a music instructor and a machinist. After moving to St. Petersburg, Florida, in 1963, he was hired by the Universal Machine Company to build aluminum boxes to house electronic components for NASA rockets. The usual process of fabricating these boxes was to cut down a 35-pound billet of solid aluminum into a box weighing between 1.5 and 3 pounds.[12]

Veleno had continued to give music lessons at home in his off-hours. According to the story Michael Wright recounted in *Guitar Stories*, Veleno wanted to put up a sign out front advertising his services, but St. Petersburg's ordinances prevented him from displaying any sign larger than one square foot. Considering his dilemma, Veleno decided to build a guitar-shaped mailbox to place at the curb outside his house. Since the mailbox had a function apart from advertising, Veleno reasoned, it could be larger than the limit specified by the ordinance and would provide him with both mail and the advertising he wanted. To build this mailbox, he would rely on the experience he had building aluminum boxes for NASA; thus, the guitar mailbox would be aluminum.[13]

John Veleno shared this plan with his aluminum supplier, who also happened to play guitar. As he described his idea, the supplier responded: "Why make just a guitar-shaped mailbox out of aluminum? Why not make a guitar out of aluminum?"[14]

Veleno agreed that this was an excellent idea, and he began to develop a prototype in 1966. Unlike Pioli's experience, when Veleno took his all-metal guitar to local nightclubs, he did not find a willing market for his innovation. He stopped work on the guitar for several years, resuming in 1970 when the same aluminum supplier asked about the guitar. Veleno agreed to show it to him, and the supplier became excited. This time, they took the guitar to a club and the guitarist on stage played it all evening.[15] The experience prompted Veleno to show the prototype to other guitarists, notably Joe Walsh of the James Gang and Jorge Santana (brother of Carlos Santana). Jorge Santana was especially intrigued, offering design suggestions that Veleno subsequently incorporated.[16]

With interest rising, Veleno created a guitar that both resembled more famous designs, like the Fender Telecaster (which had a similar body shape) and the Gibson Les Paul (which had a relatively flat finger radius on the neck), and looked distinct. The aluminum body and neck were unusual, and Veleno decided to make the headstock very pointy, resembling a bird's open beak. Veleno's experience building the hollow boxes for NASA allowed him to develop a deceptively light instrument. A standard Veleno original has a hollow body carved from two separate solid blocks (initially using alloy 7075, later switching to 6061) whittled down from 17 pounds to 1.5 pounds. The neck is thinner than that of any wooden-necked instrument, so much so that users of vintage Telecasters (known for their thick, round "baseball bat" necks) have a significant adjustment to properly finger the notes. The all-metal bodies allowed Veleno to chrome plate individual guitars or anodize them with different colors, including black, red, gold, and blue. (They also would be relatively simply to disassemble and melt down, should users wish to scrap them. It was not Veleno's intention, however, to design for disassembly, nor, for reasons discussed later in this chapter, is it likely that these guitars would ever be recycled.) Like Wandré guitars, Velenos look strikingly distinct from their wooden counterparts. Producing guitars as a one-man operation, Veleno made and sold about four dozen initially, pricing them at $600 apiece ($3,685 in 2015 dollars).[17]

The first two sales, Veleno told Michael Wright, were made at a 1972 Florida concert by the British glam-rock band T. Rex. Veleno showed T. Rex's leader, Marc Bolan, the guitar, and Bolan loved it so much that he told Veleno, "I want two, one for me and one for my friend Eric Clapton."[18]

Even then, Eric Clapton was one of the most famous guitarists in the world, and he had helped popularize the Fender Stratocaster in the 1960s. After he received his Veleno, subsequent adopters were as diverse as Dolly Parton, Todd Rundgren, and Sonny Bono. The blues guitarist Johnny Winter mused about his Veleno to *Guitar Player* in 1974, telling the magazine about "a really strange all-metal guitar made by John Veleno. It's got the thinnest neck in the world. Since it's solid metal, you don't have to worry about it warping. But I'm not quite used to it. The neck's a little too thin. The worst part about it is that the neck is silver, and its got little black dots on it, and when the spotlight is shining on the neck I really can't see the dots, so I haven't been using it on stage. But he makes pretty nice guitars. If I played it, and got used to it, I think it'd be a real nice guitar to play."[19]

Although the guitars were unusual enough to give Winter some difficulty, a larger logistical problem was scaling up production. John Veleno was building the guitars by himself or with his son Chris, and handling orders was difficult.

The distinctive "bird's head" headstock of a Veleno guitar. Photograph by Carl A. Zimring

Adding to the difficulty, many potential customers wanted modifications of the original design. Veleno did produce two units of one distinctly different model, an Ankh (featuring a minimal body with holes in it around the aluminum neck, similar in some aspects to a few of the more radical Wandré designs), at Todd Rundgren's request in 1977, as well as one unit of a bass guitar and about a dozen short-scale Traveler guitars. But Veleno produced fewer than 190 guitars in total before ceasing production in 1977.

Travis Bean

On the other side of the United States, another garage tinkerer began developing aluminum guitars. Like John Veleno, Travis Bean was a motorcycle racer and sculptor who also played guitar. He had walked into the Killeen Music Store in Burbank, California, in the early 1970s intending to purchase a small acoustic guitar. "But I found out they needed someone," he told *Guitar Player* in 1979. "I talked to the owner. He'd known me since I was a little kid, so he put me to work.

Quite honestly, I buffaloed my way in and proceeded to learn as much as I could as quickly as I could from what I'd overhear or read in old catalogs. It occurred to me that there hadn't been that much that had happened with guitars, and there wasn't that much to learn. With respect to electrics, you could know the history in a couple of weeks."[20]

Bean befriended the store's repairman, Marc McElwee:

One thing I'd learned early on was to be a good question asker and a good listener, and I proceeded to use that same approach with Marc as he supplied me with the details of guitar making. I was astonished that guitars required as much maintenance as they did and that the necks were continually going berserk. Marc was spending most of his time doing fret work or compensating for instruments that had neck problems During the first month I worked at the store, I became aware that the various instrument components—pickups, gears, and so forth—were available. Since I didn't have enough money to buy an electric, I set out to make one of my own after soliciting Marc's aid.[21]

Bean's solution to the neck problem resembled Veleno's, though the California designer claimed not to be aware of precedents.

Right off, I thought of building the neck out of aluminum. It would be easy for me to sit here and make up some story and say that I had some vast knowledge of sound and knew that a guitar was really going to perform nicely with an aluminum neck, but that wasn't the story at all. In the beginning it was done strictly from a maintenance standpoint. It was based on my *non*-experience with guitars and my ignorance of old wives' tales that would have made me believe a metal neck had to be wrong. I just kind of brazenly built it without any thought as to whether or not it was a good idea. Quite simply, I figured the aluminum neck would solve the problems that I'd been seeing Marc having to repair all the time.[22]

Bean machined the neck from a solid billet of Reynolds 6061-T6 aluminum. It was a neckthrough design running from the headstock all the way down to a "receiver section," where Bean mounted the pickups and bridge. Bean then mounted the neck to a body made of koa or magnolia wood.[23] The body, due to its thinness, needed additional reinforcement. To stabilize it, Bean installed a 3/16-inch-thick plate underneath the neck that extended back to catch the strings. Once the guitar was finished, both Bean and McElwee were "pleased with its feel, playability, and sustain."[24]

The two men decided to form a partnership under Bean's name in 1974 and began producing the guitars in Sun Valley, California. Shortly afterward, Gary Kramer, a man who had once given Bean a job at a sports car dealership, joined the business. The partnership was successful, with the first dozen guitars they produced selling immediately. This forced Bean, McElwee, and Kramer to consider whether their initial production model of a dozen guitars per month was sufficient to keep up with demand, and they soon expanded into a larger facility with new equipment. The new complex included milling capabilities, making it possible for Bean to get the necessary angles to hollow out the interior of a single piece of metal to come up with his patented one-piece neck.[25]

Unlike Veleno guitars, Travis Bean guitars mixed aluminum with wood. Bean experimented with various woods for the bodies of his designs, eventually settling on koa because of its consistency of grain. From the same 20-foot board, he could produce several distinctive instruments. He left some bodies with the natural grain intact and painted others white, black, or occasionally other colors.[26]

Although Bean's bodies were wood, the necks were aluminum, with a distinctive T-shaped hole in the headstock. "The more rigid the surface over which the string is stretched," Bean said in 1978, "the longer it will vibrate (sustain), and the less it is affected by feedback. Aluminum—first chosen because it was durable—didn't solve all the problems for us. A solid ingot of harder material also robs vibration due to its mass."[27] Bean explained the logic of the patent he filed in 1974. "That is why we've hollowed out the base of the Travis Bean neck assembly and tapered it to form a chassis for the length of the string. It is this patented chassis that makes the Travis Bean guitar what it is—an instrument that has become the most dramatic breakthrough in electric guitar technology in 50 years."[28]

Aluminum was also appealing for aesthetic reasons. "I didn't really see why a neck should be coated," he said in 1979.

> It seems to me more advantageous not to—both from a manufacturing and a maintenance standpoint. A person can put a lot of wear onto an aluminum neck—scratch it up, whatever—and within a matter of a couple of minutes, it can be buffed to "as new" condition. Moreover, you can do so a thousand times without any problem of wearing. That's not true with wooden instruments that are painted and subject to chipping. Nevertheless, we decided to

create a more orthodox-feeling surface for those who wanted one. We offer a choice of an uncoated neck or one painted with DuPont Imron—the same thing they paint airplanes with. It's about 20 times as tough as the paint we use on the body, and it offers a really nice thermal insulation as well.[29]

Bean championed aluminum as a production material, arguing, "One thing that has always fascinated me about material like this is that the improvements you make in manufacturing methods don't lead to worse instruments, they lead to better ones. That's not always the case with wood. We believe that the more sophisticated our methods and machines have become, the better our guitars have become—we can control them better and make everything more accurately."[30]

During the years of active production, Travis Bean guitars and basses enjoyed high demand and critical acclaim. Rolling Stones guitarists Keith Richards and Ronnie Wood and bassist Bill Wyman used a series of Beans between 1975 and 1979, including Richards's customized 5-string TB1000S.[31] Jerry Garcia regularly used both a TB500 and a TB1000A during the mid-1970s, and Keith Levene used a rare TB3000 Wedge model in his work with the British postpunk band Public Image Ltd. Bean was overwhelmed by the interest in his designs. "We wrote $150,000 worth of business in three days by displaying three handmade guitars" at a Chicago trade show, Bean remembered in 1999. "On the flight back, we thought, 'We're in seventh heaven; here we go!' But by the time we landed at Burbank, we'd started to realize that we didn't have a *clue* how to make that many guitars [laughs]. It took about a year and a half to figure it out."[32]

Writing in January 1976, *Creem* critic Eric Gaer called the designs a "new technology in the field of solid body electric guitars and basses," with the innovation coming from "how the sound is made and how the sound gets to the amplifier." This was accomplished by Bean's use of aluminum. "Machined aluminum is used to make up the one-piece neck assembly of the guitar forming a solid metal link between both ends of the string. This relative rigidity allows the string to continue to vibrate as long as the physics of the string itself allow, thus adding to the sustaining and harmonic properties of the instrument."[33]

Gaer appreciated how the rest of the guitar supported this innovation. "The Travis Bean humbucking pickup (more powerful than most) is mounted directly to the neck, so that the signal from string vibrations is reinforced by any vibrations from the neck itself, rather than isolated from it. The goal here is to allow the string to create the most accurate sound for as long as possible. The hardwood

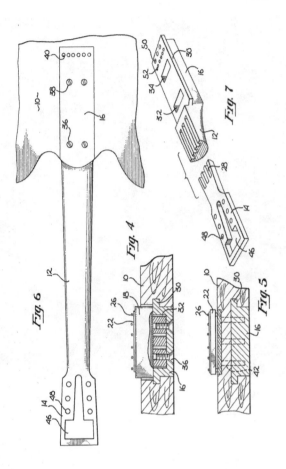

Travis Bean's patent for the neckthrough aluminum guitar design, including his signature T headstock design. US Patent 3,915,049, approved October 28, 1975

body bonded to the aluminum assembly then helps enhance the tone, as well, by mellowing it."[34]

In his review of the Bean, Gaer anticipated some of the complaints Winter had lodged against his Veleno. "Critics have registered complaints that the aluminum is subject to [the] temperature of the room and stage lights, and that it makes the instrument quite heavy. Playing the instruments, however, dispelled these objections from our thinking. The instrument is not too heavy at all (not unlike the Les

The Grateful Dead's Jerry Garcia playing a Travis Bean 1000A in 1976. Courtesy and copyright © Ed Perlstein

Paul models from Gibson), and the neck does not change appreciably in feel even under hot stage lights."[35] "The Travis Bean instruments score high marks with us," Gaer concluded. "Do a lot of comparing on your next trip to the music store." He warned, however, that "you may have to wait a while to see a Travis Bean in that current orders heavily outweigh their ability to produce. They're catching up though and should have enough instruments out by Christmas time to give you a good, long look-see."[36]

In November of the same year, Gaer's *Creem* colleague Allen Hester reevaluated Travis Bean guitars, concluding:

The simple fact is this: the precision demanded in recording and in concert is better met by the Travis Bean, and many significant players have already realized this and embraced the instrument as a genuine innovation. The guitars stay in tune better and the pickups deliver clear low distortion tone at both extremes in volume. Furthermore, the dense, highly figured Hawaiian Koa wood body reinforces sustain and enhances resonance. The hand-rubbed lacquer finishes are handsome without being gaudy, particularly the hand-carved top of the TB1000 Artist model. The flat rosewood finger-board and jumbo frets make fingering much easier, especially string-bending and vibrato.[37]

Hester noted that the TB1000 Artist retailed for $699 ($2,911 in 2015 dollars) and said that was a reasonable price for such a fine instrument.[38]

In 2005, *Guitar Player* columnist Art Thompson fondly recalled Travis Bean's active period: "The polished aluminum neck was a key part of Bean's visual esthetic. I remember seeing Jerry Garcia during his Travis Bean phase, and the reflections the guitar radiated to all points in the hall were pretty spectacular, as were the bright, bronzy tones it produced through his Fender/Macintosh/JBL rig. The Bean's pristine, hi-fi response was obviously well suited for Garcia's style, and in a rare endorsement, he declared, 'The best damn production guitars and basses in the world are made by Travis Bean.'"[39]

Despite the quality and the recognition from famous musicians, Thompson noted economic drawbacks to the guitars. A 1978 price list had the model 500 at $500, the TB1000S at $995, and the TB1000A at $1,195. Left-handed models added another $200 to the price. (In 2015 dollars, that is $1,817, $3,617, and $4,344, respectively, with an increase of $727 for left-handed models.) By 1979, Thompson remembered, "the company's investors began calling for the prices to be lowered. Not willing to cut corners and diminish the quality, Bean chose instead to stop production."[40]

In 1999, Bean recalled the logistical difficulties:

Without a history, banks wouldn't lend us money, but they *would* lend us money to buy *machinery*. We were able to put together a wood shop, a paint shop, and a metal shop. I had a tremendous bunch of folks working with me, not *for* me, and we managed to make about 3,000 instruments. I try to give credit where credit was due; my machinist really figured all of the tooling out. We looked into casting and forging, but it was so outrageously expensive for the dies that we ended up doing it the old-fashioned way.

We had 21 people working during those five years. Only one person left; we knew we were working on a good product.[41]

Although Travis Bean's designs employed more wood than John Veleno's, and Bean was able to produce thousands of guitars and basses (compared to fewer than 200 Velenos), both designers were plagued by issues of scale.[42] Both created coveted and lauded designs used by famous contemporary musicians. Both had small production operations with a handful of employees, resulting in high-quality instruments but low numbers. The waiting times for musicians who ordered these guitars grew longer, and the companies could not keep up.

Kramer

Despite Eric Gaer's optimism that Bean's capacity would catch up to demand, that did not materialize. Bean's business was hampered by its own growth in that financing and hiring were difficult for a business that could not produce enough guitars to sell to raise the capital necessary to complete production on all of its back orders. Disagreements over capitalizing the company spurred Gary Kramer to split off and create his own aluminum-necked guitar company, working with partner Henry Vaccaro.

Before Bean's active career ended in 1979, he wished his former partner well. "I would much rather Kramer guitars be successful than fail," Bean said. "I feel the more people involved in improving guitars in general the healthier it is for everyone."[43]

Kramer guitars did succeed, in that the company proved to be the longest-lived of the 1970s manufacturers that produced aluminum guitars. But it eventually abandoned the aluminum models of its initial focus in 1976 to work with more conventional woods. For most of its first decade, however, Kramer produced a variety of aluminum-necked guitars and basses, taking over from Bean as the most prolific of the aluminum luthiers.

When the company began, Kramer commissioned the New Jersey–based luthier Philip Petillo as a consultant designer and engineer. Petillo handcrafted the prototype designs and assisted in the setting up of a production line. Unlike the neckthrough Bean design, the Kramer design was based around a solid forged aluminum neck that was bolted to a traditional solid wooden body. Kramer's Dennis Berardi designed a neck based on a forged aluminum T section that had shaped curly maple wood on the back and an ebonol synthetic fingerboard bonded to the neck. In this way, the Kramer player's hands were shielded from the aluminum while the metal gave the neck the strength and rigidity of a Bean or Veleno.[44]

In 2005, Art Thompson recounted that after Travis Bean ceased production, the "metal concept would stick around for years via his ex-partner's line of Kramer-brand aluminum-necked guitars. The caveat, of course, was that Kramer's design incorporated wood inserts that were set into the neck to provide a more natural feel—a detail one might conclude was a harbinger of wood's ultimate triumph in the great metal challenge."[45]

Regularly produced Kramer models in the late 1970s included the 450G (guitar) and 450B (bass); the cheaper 350G, 350B, 250G, and 250B models; the DMZ series; and the XL series, which began the company's transition to unconventional body shapes intended to appeal to the growing heavy metal market. Kramer also produced specialty models, including a 1981 bass guitar shaped like an ax for Gene Simmons of Kiss.[46]

That market became Kramer's focus in the 1980s. To cater to heavy metal players, the guitars featured hot pickups and, after 1981, wooden necks instead of aluminum ones. Kramer was the most public face of aluminum instruments by the early 1980s, but the company discontinued aluminum-necked production entirely in 1985. In the short run, the move away from aluminum benefited Kramer, since the market for flamboyantly designed wooden guitars soared during the rise of heavy metal in the mid-1980s. But the company faced serious financial problems by 1991, entering bankruptcy and ultimately having its assets purchased by the Gibson Guitar Corporation. Founder Kramer left, eventually starting Gary Kramer Guitar in 2005. Founding partner Henry Vaccaro purchased the original Kramer aluminum design patent, and the Vaccaro Guitar Company produced small numbers of new aluminum-necked guitars between 1997 and 2002. Perhaps the most famous musician to play these guitars was U2's the Edge.[47]

Aluminum Amplifiers: Hartke Systems

As Kramer saw the end of the production of aluminum guitars, aluminum found a new use indirectly relating to electric guitars. The amplifier manufacturer Hartke Systems saw aluminum as an improvement over conventional speaker construction. Most amplifiers featured paper cones that could tear when traumatized. Hartke's innovation was to replace the paper with aluminum, reasoning that the metal would be less fragile than paper, yet still flexible enough to perform as paper would. The jazz fusion bassist Jaco Pastorius was an early adopter of Hartke's amplifiers, and the designs have remained in production. Hartke's successful niche in the amplifier market stands in contrast to the decline of aluminum guitars.

The Destruction of Value

The history of aluminum guitars reflects Michael Thompson's dynamic model of value, especially in that the value of these instruments was socially informed.

Kramer ceased production when aluminum guitars were not in demand. By the mid-1980s, the instruments were historical artifacts found in pawnshops, their economic value diminished. Most performed just as well as musical instruments as they had when they were new, but stylistic obsolescence had reduced their value.

Some complaints about aluminum guitars were similar to Johnny Winter's 1974 concerns about thin necks and unusual feel. In 1978, guitar designer Donald Brosnac remarked: "The major drawback the author finds with metal neck guitars is that the neck feels cold and cramps his hand."[48] In a 1988 patent filing for a "body for an electronic stringed instrument," the inventor Eric Clough critiqued the designs of Bean and Kramer. "The Travis Bean guitar," Clough argued, "is disliked because it is expensive, very heavy (20–25 lbs.), and requires a casting almost three feet long." After exaggerating the Bean's weight by a factor of two (weight that in large part was due to the wooden portions of the instrument), Clough criticized the Kramer's bulk. The Kramer bolt-on neck "was supposed to provide the same attempt to increase sustain in a less costly, lighter format," but it "is still heavy and it still requires a relatively large casting."[49]

Aesthetic changes in some musical genres led to a new embrace of aluminum guitars at the end of the 1980s. Since then, aluminum-necked guitars have been favored by performers in genres as diverse as country music (Buddy Miller), jazz (Stanley Jordan), and punk (Steve Albini, Duane Denison, Lee Ranaldo).[50]

New custom-built aluminum guitars emerged. One high-profile example came from Chicago-based Ian Schneller's Specimen Products. The Illinois-based rock band Tar's guitarist John Mohr requested from Specimen an "indestructible guitar." Schneller responded with a bespoke body made of aluminum (and, reversing Bean's and Veleno's designs, a wooden neck). Tar featured the Specimen guitar on the cover of its 1993 album *Clincher*; Mohr used the same guitar in reunion shows 19 years later.[51]

The new Specimen guitars contributed to the appreciation of existing aluminum guitars among some rock musicians, especially those making aggressive punk, noise, drone, or heavy metal music, with adopters, including Stephen O'Malley of Sunn O))) and Steve Albini of Shellac.[52] Albini had played a TB500 in the 1970s, and after purchasing one in 1990, the guitar became his primary instrument. Albini remarked, "I have since become very fond of certain aspects of this guitar, in particular the way the neck bends easily for warping the notes but doesn't go out of tune on its own. I often grab the T headstock with my free hand as a handle for bending."[53] This style would break many wooden guitars, but

Tar's *Clincher* album cover, 1993. Original photograph by Bob Hansen; rights and reproduction courtesy Touch and Go Records

"aluminum guitars are durable and hard to break." Of Velenos and Beans, Albini remarked: "The instruments themselves lasted longer than the business instrument behind them."[54]

Albini's enthusiasm for his TB500 led him to suggest in 1992 that Silkworm bassist Tim Midyett try a Travis Bean. Midyett, after playing Albini's guitar, was impressed with "how lively it was. It was exciting to feel it respond so directly to the physical input when I picked it." Subsequently, Midyett scoured the used guitar market and bought a used TB2000 bass for $300 ($506 in 2015 dollars). Midyett made the bass his primary instrument because "the sustain and evenness of physical response are big positives. It's not trivially easy to find a [Fender, wooden-necked] P[recision] bass with no dead spots on the neck—my Wedge has been like that since the day I bought it off the wall of a malt shop. Also the frequency response [is] very even and wide. It's more like the frequency response of a piano than a typical electric guitar."[55]

Tim Becker first became aware of aluminum guitars when he heard Tar's *Clincher* album in the early 1990s and was reacquainted with aluminum in 1997 when he saw a Travis Bean for sale in a guitar shop while on tour with his band Smeller. "I was on tour in Bellingham, Washington, and wandered into a music store in a mall and saw a TB1000—serial number 666." Its price, Becker remembered, "was five or six hundred dollars. I stupidly didn't buy it, and when I went back several months later it was gone." Becker later purchased a TB2000 bass guitar.[56]

Guitarist Jodi Shapiro has performed with the avant-garde composers Glenn Branca and Rhys Chatham. She has played Beans for two decades because a Bean "can withstand all the whacked-out tunings that Branca and Chatham throw at me (and it), and I don't have to worry about uneven string tension warping the hell out of it. Along those same lines, I can do stuff to them that might damage a wooden neck. I often tap on the back of the neck, hit it with bottles or other objects to make sounds."[57]

Shapiro acquired two Travis Beans when the guitars were near the nadir of their value. In search of a Bean in 1995, she called a Canarsie, Brooklyn, music store. The store had a TB1000, so she went in and "looked it over with my non-expert eye. It looked fine to me, so I asked how much they wanted for it. The guy looked at me, sized me up, and said '$250.' I looked him square in the eye and said, '$200 plus a strap. We both know this is going to be sitting here for months if I don't buy it today.'" He agreed. The following year, Shapiro spent $300 on a TB500 she saw on a Seattle music store's website, bringing her collection to two. (Shapiro paid the equivalent of $311 in 2015 dollars for the TB1000 and $453 in 2015 dollars for the TB500.) Two decades after the second purchase, she still owns both guitars. Both have appreciated in value to the point that they are worth thousands of dollars, and Shapiro says she has turned down an offer of $7,000 for her TB500.[58]

Shapiro's anecdote is supported by price guides for used guitars. The vintage guitar market in general increased after 1990, bolstered by an emerging collector's market and by interest among professional musicians. Some of these guitars gained value due to the celebrity of their previous owners (such as the $312,000 Garcia-owned Travis Bean), but surveys of the annual price guides from *Vintage Guitar* magazine reveal rising prices for aluminum-necked guitars irrespective of such attachment to famous musicians.[59] In the 2007 guide, for example, most Travis Bean guitars and basses produced during the 1970s sold for between $2,000 and $3,000 ($2,286 and $3,429 in 2015 dollars), with the rare Wedge guitars and basses selling for between $3,000 and $4,000 ($3,429 and $4,572 in 2015 dollars). In the 2014 guide, the vintage Travis Beans sold for between $4,500 and $6,000, with the Wedge models selling for between $4,700 and $5,800.[60]

Estimated 2007 prices for Wandré guitars ranged between $1,500 and $4,000 ($1,714 and $4,572 in 2015 dollars). In 2014, Wandrés sold for between $2,200 and $6,000.[61] Estimated 2007 prices for Veleno guitars reflected their relative scarcity. The original model sold for between $8,000 and $9,000 ($9,144 and $10,288 in 2015 dollars) with the rare Traveler model selling for between $10,000 and $13,000 ($11,431 and $14,860 in 2015 dollars). The 2014 price guide only estimated the value of the original model, pricing it at between $4,000 and $9,000.[62]

Kramer guitars and basses were more affordable than the other vintage aluminum-necked guitars, reflecting their relative abundance in the marketplace. Estimated 2007 prices for a variety of Kramer instruments with aluminum necks produced between 1976 and 1981 ranged from $350 to $900 ($400 and $1,028 in 2015 dollars), with most priced between $500 and $700 ($571 and $800 in 2015 dollars). Fewer Kramer models were listed in the 2014 guide; those that appeared ranged in value from $700 to $1,325 apiece.[63]

The growing value of these guitars over the past quarter century is in part due to the stewardship of their 1990s owners. In addition to maintaining and playing them, aluminum guitar enthusiasts began discussing their instruments on emerging discussion lists and websites. Shapiro developed the first online Travis Bean database, where owners could post their specific make and model number; multiple sites have followed suit. Since 2001, Hank Donovan's Travis Bean Guitars: Unofficial Guitar Resource has provided public discussions, photographs, videos, and a frequently updated database with information on more than a thousand individual guitars from Travis Bean's 1974–1979 production run. A similar site, MetalNecks, hosts discussions of all makes and models of aluminum-necked guitars.[64] Donovan is developing a documentary film about Travis Bean; time will tell if this raises the profile (and value) of these instruments even further.

The durability Shapiro values in her experimental music is also attractive to guitarists who make aggressive rock music. Steve Albini's technique with his TB500 regularly includes bending the neck and otherwise performing what would be abusive to a wooden guitar—including behavior that tests the limits of the wood-aluminum Travis Bean designs' durability. During a Shellac concert at

Detroit's State Theatre, Albini managed to break the wooden body of his guitar. "I broke it by acting like an asshole. I was whacking my amplifier with the headstock, and the body split down the middle."[65] As a result, "the whole middle of it popped out about an inch, making three pieces just barely held together by the pickguard and a couple of wood fibers."[66] The integrity of the neck was uncompromised, however, allowing for a successful repair. "The repair was a glue-up and a metal plate . . . making it essentially indestructible."[67]

The instruments Bean, Veleno, and Pioli made have appreciated greatly in value. A Travis Bean guitar once cost about $400, and today the guitars sell on eBay for more than $3,000. (A September 2013 search of eBay revealed Travis Bean models for sale at $3,200, $3,395, $4,250, and $4,400.) While scarcity and "antique" pricing may account for some of the appreciation of these now-discontinued instruments, the perceived performance of these forty-year-old guitars is a crucial element in their current value.

New Production: Obstructures and Electrical Guitar Company

In 1997 and 1998, Travis Bean briefly produced new versions of his 1970s guitars under the name Travis Bean Designs.[68] This new activity did not last beyond producing a couple of dozen instruments, but other designers also began producing new aluminum instruments. In 1999, the band New Brutalism developed the ABC Group Design + Documentation collective to create aluminum instruments. Since then, the group (now named the Obstructures design collective, led by Matt Hall of Auburn, Alabama; Brian Johnson of Knoxville, Tennessee; and Nathan Matteson of Chicago) has created about 20 aluminum guitars, bass guitars, and drum shells for clients. One promotional image shows an automobile running over an aluminum guitar, which survives the ordeal without damage. In addition to New Brutalism, the band Oxes uses Obstructures instruments.[69]

Kevin Burkett has become the most prolific twenty-first-century designer of aluminum instruments. Burkett is a musician whose band Gravity Keeps the Hours recorded a session in Atlanta with Steve Albini in 2003. Burkett and Albini began discussing their mutual appreciation of Beans and Velenos, leading Burkett to work on merging favored aspects of each into his own designs. During a discussion thread about Travis Bean guitars on the Electrical Audio forum in November 2003, Burkett announced that he was building four prototypes (three guitars and one bass) in consultation with Albini and asked the forum if "any of you [would] be interested in a solid aluminum guitar and bass that combines all the wonderful features of a Veleno and Travis Bean."[70] Finding favor for his idea, he became an apprentice at a machine shop in Pensacola, Florida, working with Performance Machining's Will Fitzpatrick on the prototypes. They produced about 10 instruments in the first year, and work on EGC instruments eventually took enough of Burkett's time that he left Performance Machining to establish his own shop in Pensacola's W Street Industrial Park. This was followed by a move to a larger space in the same park in 2009 to handle increased demand.[71]

Like Albini and Shapiro, Burkett discovered aluminum-necked instruments in the early 1990s when he needed a bass guitar and discovered Beans.[72] Influenced by both Travis Bean's and John Veleno's designs, Burkett took a cue from Steve Albini's repair to his TB500. "The EGC500 was conceived for Steve Albini. He is an avid TB500 player, so we set out to make a guitar that would withstand the abuse that the original (nor any guitar, really) was never intended to endure. We started with an alder body and our core EGC neck through and added a period-correct aluminum chrome plated bridge receiver with chrome plated brass saddles. As Steve had done for his TB500 after he broke it in half, we anchored the entire guitar to a massive aluminum resonator plate, which made it essentially unbreakable."[73]

Burkett's designs take elements from both Bean and Veleno designs. Like the Beans, the EGC has a neckthrough design attached to the body (which may be wood, aluminum, or Lucite) and a headstock with a signature hole. Like the Velenos, the EGC has a very thin neck profile. Burkett has a series of standard guitars, but also (in contrast to Veleno and Bean) has worked with many clients on custom designs ranging from 12-string all-aluminum bass guitars for Night Mode's Chris Hall and Cheap Trick's Tom Petersson to hollow-bodied and solid-bodied baritone guitars (several for Tim Midyett) to a Lucite-bodied Flying V guitar for Brent Hinds of Mastodon. "I started doing all the custom stuff so all my research and development was done by others' ideas. The core design was there, but it was good to try so many combinations."[74]

Burkett's shop provides him the capacity to both fill orders beyond what Veleno could achieve and provide a range of options, including offering guitars designed by Uzeda's Agostino Tilotta, the Jesus Lizard's Duane Denison, and Shellac's Todd Trainer. "It's a full machine shop and wood shop so we can make whatever," but Burkett prefers working with aluminum, calling it "an amazing format, and I wonder why it's not used more often. I dig wood, but it will eventually rot away."[75]

Like Veleno guitars, Burkett's all-aluminum models are highly durable. Much or all of the instrument could be recycled if the user wished to do so, yet the guitar's enduring functionality in its manufactured state renders discussion of

Tim Midyett playing his EGC baritone guitar and Brian Orchard playing bass guitar at a Bottomless Pit show, July 11, 2008. Photograph by Jodi Shapiro

disassembly moot. Burkett's standard models and his willingness to work with the design preferences of customers have attracted guitarists who used Beans and Velenos as well as others who are new to aluminum guitars. Fellow guitar designer Earnie Bailey said of Burkett: "He's taken the best of the two companies that came before him and made the perfect instrument out of the two of them."[76]

Early in Burkett's production career, he contacted Tim Midyett for design ideas for a baritone guitar. "He asked if I had any advice," Midyett recalled in 2013.

> I told him what I thought about scale length and pickups and neck profile and bridge and stuff. Then he asked if I'd play one if he built it. I told him well, you know, I really love my Travis Bean that I have modified into a baritone. So I probably wouldn't be interested. And he said he'd make one and send it along so I could try it out, no obligation.
>
> I got his prototype, played it for half an hour. Played my Bean for 30 seconds, called Kevin, and offered it to him in trade.
>
> He really thinks about how they sound, in detail. He could have easily kept putting other people's (very good) pickups in his things, but he wasn't quite happy with what he could get, so he started making his own. He just has great

attention to detail without losing sight of the overall reason he's doing it, which is to give people something that truly signifies in terms of sound.[77]

Guitarist Buzz Osborne of the Melvins was a regular Gibson Les Paul player until 2010, when he saw the EGC guitars the metal band Isis used. "We rehearsed at the same facility in L.A. as they did, and I saw these weird aluminum guitars. I picked one up, and what instantly sold me on it was the neck. It was the same thickness from the headstock to the body. You can't get that thin of a neck with wood because it'll break. Plus, I have the hands of a five year old. You look at a picture of [Jimi] Hendrix and he could wrap his finger around a Strat neck like four times, whereas a Strat neck looks like a Precision Bass in my hands."[78] Osborne called Burkett and asked if he could make an aluminum guitar with a Les Paul scale. "And he did it. What people don't realize is that these aluminum guitars actually have more low end than Les Pauls. They're wonderful guitars. I have about seven of them."[79]

Chris Rasmussen, who plays bass in the rock band Police Teeth, owns an Electrical Guitar Company bass because it is "aggressive with good high end, but also very clean and with the sustain I'd expected. It worked great for my playing style. I especially loved how hitting harmonics would be way more resonant than with a wooden-neck instrument, and how easy it was to get a ringing note to begin to feedback and sustain."[80]

Testimonials such as Osborne's and Rasmussen's have led to higher demand for EGCs, and musicians such as Cheap Trick guitarist Rick Nielsen (who purchased Travis Bean guitars in the 1970s) are among the clients ordering the new aluminum guitars. In 2011, Burkett noted that EGC had produced 522 instruments and had a waiting list of 122 orders to fill.[81] The waiting time for Electrical Guitar Company models was more than a year, reflecting Midyett's sentiment that they are "the finest production guitars in the world."[82]

Sustained Value

Between 2010 and 2015, instruments produced by Obstructures and the Electrical Guitar Company sold for between $1,200 and $3,500 new, hundreds of times the value of the weight of scrap aluminum used in their manufacture. Prices depend on the quality of the materials used (the aviation-grade alloys 6061 and 7075 are Kevin Burkett's choice for his guitar necks), the aesthetics and performance of the finished instrument, and rising demand, which requires expanded investments in production. Unlike Travis Bean's challenges in the late

1970s, EGC production has expanded without sacrificing quality or encountering serious economic problems. The result has been more than a decade of expanding production.[83]

These instruments could, if needed, be melted down and recycled. Their present value and presumed durability (based on the history of the older aluminum guitars) make that fate unlikely for most of them. The Oakland, California, guitarist Conan Neutron's experience with the Electrical Guitar Company is instructive. In 2012, he owned one guitar, an old Squier Jazzmaster with a wooden neck, which he estimated to be worth $300. He played in a band with a guitarist who owned an EGC and was both intrigued by its beauty and concerned about the performance and weight of the instrument. With that in mind, he contacted Burkett with a request to "make magic" with a custom guitar—a nice, aluminum version of his Squier Jazzmaster that would perform like the EGC he had heard without the weight concerns. Burkett agreed to develop the custom design while Neutron sent a down payment followed by subsequent installments.[84] Although this was during a time when the demand for standard and custom orders had increased, the Electrical Guitar Company was able to deliver Neutron's custom model in early 2014. Modifications to the standard design, including a thin neck and a hollow body, eliminated the weight concern. Other tweaks to the original design included a pickguard inspired by the bespoke Specimen aluminum guitars created for Tar's John Mohr.

The efforts of designing and paying for this instrument reflect the value of highly coveted durable goods. Neutron now owns two guitars: the Squier and the EGC. After spending some time getting used to the thin neck profile and the tactile sensations of the aluminum neck, Neutron spends most of his playing time on the EGC. It feels more comfortable to him and produces more satisfying music, in part due to the extended sustain of notes. Because the neck does not bow, it can hold unusual tunings well, which allowed him to use a drop C tuning to write and perform on an album. For the owner, the guitar is both a thing of aesthetic beauty and, for Neutron's purposes, superior as a musical instrument. He termed the EGC "elegantly utilitarian," and noted that he plays far more often now that he has this guitar.[85]

In contrast to the McDonough and Braungart vision of upcycling industrial materials as technical nutrients, the users of Wandré, Bean, Kramer, Veleno, and EGC guitars see the aluminum instruments much more like Charles Eames saw his chairs—as objects to covet. Some aficionados collect or trade the guitars. Rick Nielsen, for example, owns Beans as part of his collection of more than 400 guitars; the actor Vincent Gallo has spent several years amassing a collection of Velenos and Beans.[86] Like paintings and other artworks, these guitars now have cachet as luxury items.

Yet aluminum guitars are also utilitarian and are coveted by users who are not collectors. Neutron only owns one EGC and in 2015 declared that he "will never sell" the guitar. When he is not playing it, it is displayed on a wall at his home. It has an intrinsic emotional value for him that would preclude disassembling it and returning it as raw material for industrial production. The durability is part of the aesthetic appeal; Neutron appreciates that the guitar is made of the same material as airplanes.[87]

Time will tell if Neutron changes his mind about selling, but if he keeps the guitar, he has precedents in Buddy Miller (who has owned his Wandrés for 40 years), Steve Albini (who has owned his TB500 for a quarter century), and Jodi Shapiro (who has owned her two Beans for 20 years). The track record of EGCs as durable goods in the vein of past aluminum guitars has led to a resumption of production of Bean designs. In 2013, the Electrical Guitar Company began manufacturing versions of Bean's 1997 designs, licensed from his widow, Rita Bean. Some of the new guitars feature the same shapes as the 1970s models, but with a stabilizing aluminum pan on the body. Like the Eames Aluminum Group furniture still being produced by Herman Miller, Bean guitar designs have enduring appeal. Also like the current versions of classic Eames Aluminum Group designs, the new Travis Bean Designs series is expensive. New models sell for $7,500 apiece.[88]

All of these guitars are potentially recyclable, although the level of processing required for harvesting the aluminum depends on the individual design decisions to use wood, paint, adhesives, and coatings, which would need to be separated. The designs of the 1950s, the 1970s, and the 2010s all feature materials that are reasonably easy to recycle. The history of the use of these guitars, however, works against these products being recycled. Despite periods when the instruments lost cachet and saw their prices decline, Veleno and Travis Bean designs from the 1970s have experienced a sustained growth in price. Owners of instruments that are more than 40 years old are now more likely to repair and maintain the aging equipment than to return the materials to industrial production. The culture of guitar ownership among collectors and working musicians amplifies the ethos of durable goods at the same time that it discourages treating the instruments as technical nutrients for further production.

Scrap aluminum contributes to the creation of new EGC and Obstructures guitars, but the likelihood that the scrap in new guitars comes from the Beans and Velenos made in the 1970s is remote. The value of the older designs exceeded the metal's value as scrap even in the nadir of aluminum guitars' popularity. After two decades of rising value, the old guitars have achieved vintage status. Newer models are considered prestigious and sell for thousands of dollars. Owners of aluminum guitars old and new are aghast at the idea of scrapping them. The history of their use and trading indicates the value of the designs and is a victory for creating valued objects from reclaimed material. This history also is a lesson for advocates of the circular economy: durable, covetable goods made well will leave the circle.

Designing for Sustainability

Waste is a product of design, and design can salvage waste. Design *has* salvaged waste and created valuable goods, as this history of aluminum use has indicated. Recycling occurs because of economic and political factors. These include the market rate for salvaged material, the cost of reclamation versus disposal, and public policies to discourage waste, such as extended producer responsibility regulations. Design plays a large role in determining whether material is wasted or recycled.

Redesigning industrial production in order to salvage scrapped materials and turn them into goods of durable value appears to support the transition to a circular economy. The examples in this book show how aluminum upcycling has successfully produced covetable, durable goods. At first glance, upcycling appears to be a logical design solution to the unsustainable resource use of industrial societies. If discarded matter can be transformed into durable objects of great economic value, waste will be reduced and recycling will increase. The power of intentionality, McDonough and Braungart's argument goes, will eliminate waste and "create a more abundant, joyful world for future generations."[1]

The historical record of aluminum upcycling both confirms and complicates this picture. For more than half a century, designers have used secondary aluminum to create durable objects that retain and even increase their value with time. The history of aluminum vehicles, furniture, and musical instruments provides a basis for the claim that designers can work to close loops of material flows in industrial economies. Moreover, the fact that the majority of aluminum created in the twentieth century remains in use today provides optimistic evidence for the maintenance of industrial loops on a large scale. The durable goods detailed in this book, in one sense, are triumphant historical examples of upcycling.

The complications, however, are numerous. First, the processing necessary to recycle aluminum involves energy and toxic emissions. Some of these emissions are due to the materiality of the aluminum alloys. Others are due to the creation of goods that were not designed for disassembly. Some of these products, like the hybrid aluminum-plastic condiment packets, are simply too difficult to collect and process for the value of the material and are lost to waste. Others produce waste in the recycling process because of design issues. The aluminum laptops that Steve Jobs boasted were so green have negative environmental consequences when they are recycled. While a sleek, thin, aluminum-bodied laptop seems perfect for recycling, the design of the computer is so dense that scrap aluminum processors have to shred the entire machine in order to harvest the aluminum. Every other part of the computer is turned into shredder residue, small particulate matter that can have consequences to human and ecological health. Recycling the aluminum in such products is possible, but the process challenges the notion of recycling as a practice that has sustainable benefits to the planet's ecology and vulnerable peoples.[2]

Samantha MacBride identified a problem with C2C design in that it assumes manufacturers will voluntarily develop designs that minimize waste. But industrial history, including Apple's greenest laptops, indicates that is not the case. Voluntary design absent regulations on waste that require producers to bear the economic burdens for the disposal and recycling of those goods begets more waste.[3]

Apple's laptops share their waste issues with the Ford F-150. Once the "most sustainable truck" ever produced is wrecked or deemed unfit to drive, it, like the millions of automobiles made before it, will enter automobile shredders that separate the aluminum, ferrous metals, and other valuable recyclables from every other part of the vehicle. The F-150 will thus generate toxic automobile shredder residue, which places the waste management burdens on end-of-life processors.

Ford's signature truck raises a more troubling question for the notion of upcycling technical nutrients on a large scale, allowing abundance without ecological cost. In its early days on the market, the truck was a remarkable success, with Ford selling as many aluminum-bodied F-150s in two months as the NSX (the most widely manufactured aluminum car) sold in 15 years. This success shows that a mass market for secondary aluminum exists—and it also shows that the market for primary aluminum continues to grow.

The inconvenient truth of aluminum's history is that while recycling rates have remained high, primary aluminum production has expanded. If aluminum production's expansion from the beginning of World War II to 1980 fueled the variety of goods discussed in this book, what should be made of the subsequent 40-year period? According to data collected by the International Aluminium Institute, global aluminum production exceeded 15 million metric tons in 1980. It rose steadily over the decade before plateauing at 19 million metric tons between 1989 and 1995, rising to 20.8 million metric tons in 1996, 24.3 million metric tons in 2001, 28 million metric tons in 2003, and exceeding 30 million metric tons for the first time in 2005. Production has continued to rise, hitting 41 million metric tons in 2010 and 45 million metric tons in 2012. Despite high recycling rates, despite sophisticated public and private collection systems in much of the industrialized world, global primary aluminum production has become even more intensive; it was more than 50 million metric tons in 2014.[4]

In 2000, the United States imported 3 million tons of bauxite and 400,000 tons of alumina from Jamaica, almost all of which was used for primary aluminum. Although the vast majority of all aluminum produced remains in use, the ways that it is used invite further aluminum production.[5]

Domestic primary aluminum production in the United States has declined, but global production has increased. Mines in the Caribbean, Africa, and Australia have seen more intensive excavations in the past 40 years, and smelters in those regions have replaced smelters in North America and Western Europe. Reynolds Metal, for example, entered into a joint partnership with the government of Guinea in 1997 to operate the Aluminum Company of Guinea. The government also, with an international consortium including Alcoa and Alcan, operates the Compagnie des Bauxites de Guinée. These holdings have made Guinea a global leader in bauxite mining and aluminum production; by 2012, the nation was the second-largest bauxite producer in the world. As with much heavy industry, primary aluminum production feeding American markets has not declined. Production has simply moved beyond the nation's borders.[6]

Since the turn of the twenty-first century, and especially since 2010, primary aluminum production in China has spurred a new boom in the global market. Mining and primary production has increased despite aluminum's vast utility

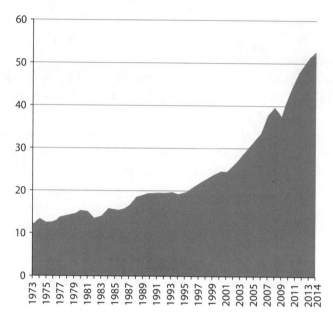

Global Primary Aluminum Production, 1973–2014 (in million metric tons). Derived from International Aluminium Institute data, compiled from voluntary reports of member and nonmember companies around the world. http://www.world-alumin ium.org/statistics/primary-aluminium-production/#data (accessed 21 December 2015)

in recycling and upcycling. In the first half of 2015, China accounted for 56 percent of global primary aluminum production, manufacturing the metal for both domestic consumption and exports throughout the world. Reuters, using statistics from the China Nonferrous Metals Industry Association, reported that the expansion of smelting capacity in the northwestern province of Xinjiang shows how dramatically Chinese production has grown. Xinjiang produced 64,000 tons of aluminum in 2008 and 4.8 million tons in 2014. The latter figure for this one province surpassed all North American production in 2014. Chinese smelters now draw on bauxite mines in Indonesia and Malaysia; primary aluminum production in 2015 was more globalized and more intensive than at any point in the metal's history. Although Chinese companies finished 2015 by announcing cutbacks in production, more primary aluminum is available worldwide at this book's writing than has ever existed before.[7]

Much of this increase may be traced to the expanded production of disposable packaging, which relates to the expansion of sustainable design strategies I described in the introduction. A rationale for upcycling is that it is a solution to reclaim waste from the land and water. If adidas reclaims plastic from the oceans to make new shoes, is that not a design strategy that helps the environment? The flaw in this analysis is that upcycling bears some responsibility for increased production and consumption. The durable covetables discussed in this book increase demand for aluminum goods. While a Norman Foster chair made entirely from recycled aluminum will have a life-cycle assessment of a closed materials loop, it helps build a market for more aluminum furniture.

As designers create attractive goods from aluminum, bauxite mines across the planet intensify their extraction of ore at lasting cost to the people, plants, animals, air, land, and water of the local areas. *Upcycling absent a cap on primary material extraction does not close industrial loops so much as it fuels environmental exploitation.* Despite aluminum's high recycling rate, the appropriate schematic for aluminum use looks less like a closed loop than it resembles an upward spiral or funnel cloud drawing more primary material into the cycle of production as material demand increases.

Secondary aluminum supplements the large technological systems that produce primary aluminum, but the relationship between mined ore, massive amounts of energy, and the environmental damage they produce remains central to the metal's use. Covetable, durable goods using the metal fuel demands for further environmental damage. An Emeco chair or an EGC guitar will not go into a landfill; more consumers coveting Emeco chairs and EGC guitars will lead to more bauxite mining *even though the covetables themselves may be made of recycled material.* More and better uses of aluminum mean that more aluminum is required.

Those are the problems facing the project of upcycling aluminum; more problems face the metal's plastic cousins in modern production. Telling a history of upcycling plastics is possible, and Patagonia's transformation of PET bottles into clothing is an example of upcycling at industrial scale. If readers are unsettled by the story of aluminum in this book, however, imagining a more optimistic tale of upcycling plastics is difficult. The challenges of sorting and processing different aluminum alloys pale compared to the problems of sorting and processing the many materials called plastics. Polyvinyl chloride (PVC), polyethylene terephthalate (PET), and polypropylene (PP) have similar names but distinct properties and different effects on human health. Mixing them is a mistake that can render materials useless for industrial production. And these are only three of the many plastics. The recycling industry and curbside pickup programs have attempted to make sorting these varied materials possible via a classification system that separates plastics into seven different categories, but it's still an uphill battle.

The variety of materials, along with the difficulties of collecting such light products as single-use plastic shopping bags and Styrofoam packaging, contributes to recycling rates for plastics across the industrialized world being less than a tenth of their disposal rate. The Keep America Beautiful campaigns showing bottles aspiring to be benches cannot bring plastics to the recycling rate aluminum had in 1960, much less to aluminum's recycling rate today. Meanwhile, industrial production extracts more petroleum to create more plastics, much as it extracts more bauxite to develop more aluminum. The insatiable appetite of industrial society grows, producing environmental effects that defy scientists' attempts to measure their variety and scope.[8]

We know we can make beautiful and valuable goods out of discarded material. To the extent that reducing materials in landfills, incinerators, and waterways is a goal of sustainable design, the historical example of aluminum gives us hope. But the history also shows that sustainability is about more than simply closing industrial loops. Developing new, desirable uses for materials increases demand for those materials.

Designers are aware of the increasing toll of resource use, which is why a growing interpretation of Dieter Rams's principle to be environmentally friendly relates to another of his principles, the principle to be unobtrusive. The most sustainable automobile design of the twenty-first century is not the F-150 aluminum truck, the hybrid Toyota Prius, the electric Tesla, or even the Rocky Mountain Institute's ultralight hypercar prototype with its carbon-fiber body. The most sustainable automobile design of the twenty-first century is not an automobile at all, but a system to distribute transportation services. Automobile-sharing programs, such as Zipcar and IGO, and the bicycle-sharing programs of several European and North American cities distribute the services of driving to a wide clientele without the damage of mass production and disposal. If three automobiles can adequately serve a hundred people, the material effects of producing three vehicles rather than a hundred vehicles are substantially smaller. The principle of providing product service systems, in use at traditional libraries and tool libraries for years, may be expanded to other goods of widespread utility. But such a system, Philip White, Louise St. Pierre, and Steve Belletire advise in *Okala Practitioner: Integrating Ecological Design*, "requires considerable design effort," including "rethinking how the service is sold."[9]

What does this mean for the future of design? If Ford successfully designed the use of its vehicles in such a manner, the total number of F-150s produced would not require nearly the quantity of recycled aluminum of the current design, and it would also respond to the mounting demand for primary aluminum. How an automobile manufacturer might sufficiently monetize a system that drastically reduces the number of automobiles in circulation is a question beyond the scope of this book.

What is not beyond the scope of this book is considering whether such a system is feasible in the designs of covetable items. Consumers might be convinced to share within a system where the products are primarily utilitarian. One could fairly describe a pickup truck as utilitarian. But would a system of sharing covetables succeed? Would a consumer wishing to possess an Emeco chair or an Electrical Guitar Company bass opt to share the good with many people?

The notion is possible, but the long history of consumer behavior in industrial societies inspires pessimism. If possessing an Emeco chair is a sign of status, ownership, as Thorstein Veblen noted more than a century ago, is the point. This behavior does not lead to sustainable material use (and, the sociologist Juliet Schor argued, may not lead to financial decisions in the consumer's best interest), but it is an enduring part of industrial culture. The design of covetables feeds this consumer impulse, which is nourished by the large technological systems that transform energy and materials into saleable goods.[10]

The impulse to rescue wasted materials is noble, the effects of creating covetables that raise demand for materials less so. Aluminum reuse within the same large technological system that developed the voracious appetite for aluminum cannot be sustainable. The history of upcycling aluminum underscores how this contemporary design strategy has limits in changing the material world.

Notes

INTRODUCTION: **Toward a History of Upcycling**

1. Jeffrey Meikle, *American Plastic: A Cultural History* (New Brunswick, NJ: Rutgers University Press, 1995), xiv.

2. Frank Ackerman, *Why Do We Recycle? Markets, Values, and Public Policy* (Washington, DC: Island Press, 1997); Martin V. Melosi, *The Sanitary City: Urban Infrastructure in America from Colonial Times to the Present* (Baltimore: Johns Hopkins University Press, 2000), 413.

3. William McDonough and Michael Braungart, *Cradle to Cradle: Remaking the Way We Make Things* (New York: North Point, 2002), 57.

4. Alessandro De Toni, "Interview: Nathan Zhang," *Cool Hunting*, 10 December 2012. http://www.coolhunting.com/culture/brandnu-interview.php.

5. Sarah Marchant, "Interview with Industrial Designer Boris Bally," *Goedecker's Home Life*, 29 January 2014. http://www.goedekers.com/blog/interview-with-industrial-designer-boris-bally.

6. Emily Lynn Osborn, "Casting Aluminium Cooking Pots: Labour, Migration and Artisan Production in West Africa's Informal Sector, 1945–2005," *African Identities* 7, no. 3 (2009): 373–386.

7. Ann Binlot, "Adidas and Parley Team Up for Sneakers Made from Recycled Ocean Waste," *Forbes*, 30 June 2015. http://www.forbes.com/sites/abinlot/2015/06/30/adidas-and-parley-team-up-for-sneakers-made-from-recycled-ocean-waste.

8. "Ocean Plastic: Seeing Opportunity in Waste," *Parley*, 23 April 2015. http://www.parley.tv/updates/2015/4/20/ocean-plastic-seeing-opportunity-in-waste.

9. David Gianatasio, "Bottles and Cans Plead to Be Recycled in New Ads for Keep America Beautiful," *Adweek*, 19 July 2013. http://www.adweek.com/adfreak/bottles-and-cans-plead-be-recycled-new-ads-keep-america-beautiful-151303.

10. Melosi, *The Sanitary City*, 413.

11. John Tierney, "Recycling Is Garbage," *New York Times Magazine*, 30 June 1996, 24–29, 44, 48–49, 53; Susan Strasser, *Waste and Want: A Social History of Trash* (New York: Holt, 1999), 289–290.

12. Historical treatments of salvage and recycling in history include Strasser, *Waste and Want*; Melosi, *The Sanitary City*; Carl A. Zimring, *Cash for Your Trash: Scrap Recycling in America* (New Brunswick, NJ: Rutgers University Press, 2005); Peter Thorsheim, *Inventing Pollution: Coal, Smoke, and Culture in Britain since 1800* (Athens: Ohio University Press, 2006).

13. US Environmental Protection Agency, "Wastes—Resource Conservation—Common Wastes and Materials: Plastics." http://www.epa.gov/osw/conserve/materials/plastics .htm (accessed 22 August 2013).

14. Samantha MacBride, *Recycling Reconsidered: The Present Failure and Future Promise of Environmental Action in the United States* (Cambridge, MA: MIT Press, 2011), 222.

15. Elizabeth Royte, *Garbage Land: On the Secret Trail of Trash* (New York: Little, Brown, 2005); Finn Arne Jørgensen, *Making a Green Machine: The Infrastructure of Beverage Container Recycling* (New Brunswick, NJ: Rutgers University Press, 2011); Raymond G. Stokes, Roman Köster, and Stephen C. Sambrook, *The Business of Waste: Great Britain and Germany, 1945 to the Present* (New York: Cambridge University Press, 2013); Adam Minter, *Junkyard Planet: Travels in the Billion-Dollar Trash Trade* (New York: Bloomsbury, 2013); Adam Minter, "How Effective Are Apple's Recycling Programs?," *MacWorld*, 11 December 2013. http://www.macworld.com/article/2068823/how-effective-are-apples-recycling-pro grams.html.

16. Prince McLean, "Apple Details New MacBook Manufacturing Process," *Apple Insider*, 14 October 2008. http://appleinsider.com/articles/08/10/14/apple_details_new _macbook_manufacturing_process.

17. Dieter Rams, "Design by Vitsœ," December 1976. https://www.vitsoe.com/files/as sets/1000/17/VITSOE_Dieter_Rams_speech.pdf; Dieter Rams, "Omit the Unimportant," *Design Issues* 1, no. 1 (Spring 1984): 24–26.

18. Philip White, Louise St. Pierre, and Steve Belletire, *Okala Practitioner: Integrating Ecological Design* (Phoenix, AZ: Okala Team, 2013).

19. Michael Thompson, *Rubbish Theory: The Creation and Destruction of Value* (New York: Oxford University Press, 1979).

20. Jennifer Wang, "Upcycling Becomes a Treasure Trove for Green Business Ideas," *Entrepreneur Magazine*, 22 March 2011. http://www.entrepreneur.com/article/219310. The first two statistics are in the article; the September 2013 statistic was generated by the author, 28 September 2013.

21. Gunter A. Pauli, *Upsizing: The Road to Zero Emissions* (Sheffield, England: Greenleaf, 1998); Bradley Quinn, *Textile Futures: Fashion, Design, and Technology* (New York: Oxford University Press, 2010); Jason Thompson, *Playing with Books: The Art of Upcycling, Deconstructing, and Reimagining the Book* (Beverly, MA: Quarry, 2010); Helen Harle, *Create Colorful Aluminum Jewelry: Upcycle Cans into Vibrant Necklaces, Rings, Earrings, Pins, and Bracelets* (Waukesha, WI: Kalmbach, 2010); Danny Seo and Jennifer Lévy, *Upcycling: Create Beautiful Things with the Stuff You Already Have* (Philadelphia: Running Press, 2011); Tristan Manco, *Raw + Material = Art: Found, Scavenged, and Upcycled* (New York: Thames and Hudson, 2012).

22. Amy Korst, *The Zero-Waste Lifestyle: Live Well by Throwing Away Less* (Berkeley, CA: Ten Speed, 2012); Maggie Macnab, *Design by Nature: Using Universal Forms and Principles in Design* (Berkeley, CA: New Riders, 2012).

23. William McDonough and Michael Braungart, *The Upcycle: Beyond Sustainability, Designing for Abundance* (New York: North Point, 2013).

24. Amory B. Lovins, *Reinventing Fire: Bold Business Solutions for the New Energy Era* (White River Junction, VT: Chelsea Green, 2011); McDonough and Braungart, *Cradle to Cradle*.

25. Thomas E. Graedel and Braden R. Allenby, *Industrial Ecology* (Upper Saddle River, NJ: Prentice Hall, 1995); Robert U. Ayres and Leslie Ayres, *Industrial Ecology: Towards Closing the Materials Cycle* (Cheltenham, England: Elgar, 1996); Thomas E. Graedel and Braden R. Allenby, *Industrial Ecology and the Automobile* (Upper Saddle River, NJ: Prentice Hall, 1998); Braden R. Allenby, *Industrial Ecology: Policy Framework and Implementation* (Upper Saddle River, NJ: Prentice Hall, 1999); Robert U. Ayres and Leslie Ayres, eds., *A Handbook of Industrial Ecology* (Northampton, MA: Elgar, 2002); Randolph T. Hester, *Design for Ecological Democracy* (Cambridge, MA: MIT Press, 2006); Thomas E. Graedel and Braden R. Allenby, *Industrial Ecology and Sustainable Engineering* (Upper Saddle River, NJ: Prentice Hall, 2010). Index searches of the *Journal of Industrial Ecology* and *Progress in Industrial Ecology* also illuminate the evolution of the discipline.

26. Rachel Carson, *Silent Spring* (Greenwich, CT: Fawcett, 1962); Paul R. Ehrlich, *The Population Bomb* (New York: Ballantine, 1968); Barry Commoner, *The Closing Circle: Nature, Man, and Technology* (New York: Knopf, 1971); Paul Hawken, *The Ecology of Commerce: A Declaration of Sustainability* (New York: HarperCollins, 1993); Paul Hawken, Amory Lovins, and L. Hunter Lovins, *Natural Capitalism: Creating the Next Industrial Revolution* (New York: Little, Brown, 1999).

27. Histories that discuss the importance of secondary commodities to industrial society include Judith A. McGaw, *Most Wonderful Machine: Mechanization and Social Change in Berkshire Paper Making, 1801–1885* (Princeton, NJ: Princeton University Press, 1987); Thomas J. Misa, *A Nation of Steel: The Making of Modern America, 1865–1925* (Baltimore: Johns Hopkins University Press, 1995); Joshua Reno, "Your Trash Is Someone's Treasure: The Politics of Value at a Michigan Landfill," *Journal of Material Culture* 14, no. 1 (2009): 29–46; Emily Brownell, "Negotiating the New Economic Order of Waste," *Environmental History* 16, no. 2 (2011): 262–289; J. McNeill and George Vrtis, "Thrift and Waste in American History: An Ecological View," in Joshua J. Yates and James Davison Hunter, eds., *Thrift and Thriving in America: Capitalism and Moral Order from the Puritans to the Present* (New York: Oxford University Press, 2011), 508–535.

28. Julian M. Allwood and Jonathan M. Cullen, *Sustainable Materials with Both Eyes Open* (Cambridge: UIT Cambridge, 2012), 47.

29. Allwood and Cullen emphasized that this figure only accounts for the creation of the metal, not the production of goods from the metal. To recycle a beer can, the coatings and other materials have to be de-lacquered; then the can is melted with a "sweetener" of primary aluminum, cast, rolled, blanked, stamped, and coated to make a new can. Allwood and Cullen estimated that the energy required to make a new can from recycled aluminum is 26 percent of what it would be to use primary aluminum; while remaining a great savings from using primary aluminum, this indicates that the energy required in manufacturing goes beyond a simple comparison of secondary versus primary materials. Ibid., 21.

30. Eric Schatzberg, "Symbolic Culture and Technological Change: The Cultural History of Aluminum as an Industrial Material," *Enterprise and Society* 4, no. 2 (2003): 226–271. See also George David Smith, *From Monopoly to Competition: The Transformations of Alcoa, 1888–1986* (New York: Cambridge University Press, 1988); Margaret B. Graham and Bettye H. Pruitt, *R&D for Industry: A Century of Technical Innovation at Alcoa* (New York: Cambridge University Press, 1990); Brad Barham and Stephen G. Bunker, *States, Firms,*

and Raw Materials: The World Economy and Ecology of Aluminum* (Madison: University of Wisconsin Press, 1994); Matthew D. Evenden, *Fish versus Power: An Environmental History of the Fraser River* (New York: Cambridge University Press, 2004); Fathi Habashi, *Aluminum: History and Metallurgy* (Montreal: Metallurgie Extractive Quebec, 2011); Mimi Sheller, *Aluminum Dreams: The Making of Light Modernity* (Cambridge, MA: MIT Press, 2014).

31. Joel Makower, "Inside the Cradle to Cradle Institute," *GreenBiz*, 28 January 2013. http://www.greenbiz.com/blog/2013/01/28/inside-cradle-cradle-institute.

32. "Jony and Marc's (RED) Auction," 23 November 2013. http://www.sothebys.com/RED.

33. Stephanie Murg, "Marc Newson's 'Lockheed Lounge' Prototype Sells for $2.1 Million at Phillips," *FishbowlNY*, 14 May 2010. http://www.adweek.com/fishbowlny/marc-newsons-lockheed-lounge-prototype-sells-for-2-1-million-at-phillips/279398.

CHAPTER ONE: **From Scarcity to Abundance**

1. William McDonough and Michael Braungart, *Cradle to Cradle: Remaking the Way We Make Things* (New York: North Point, 2002), 138.

2. Quoted in George David Smith, *From Monopoly to Competition: The Transformations of Alcoa, 1888–1986* (New York: Cambridge University Press, 1988), 10.

3. On the historical debate over the extent of Julia Hall's contribution to the discovery, see Martha Moore Trescott, "Julia B. Hall and Aluminum," in Trescott, ed., *Dynamos and Virgins Revisited: Women and Technological Change in History* (Metuchen, NJ: Scarecrow, 1979), 149–179; and Judith A. McGaw, "Women and the History of Technology," *Signs* 7, no. 4 (1982): 798–828.

4. Geoffrey Blodgett, *Oberlin Architecture, College, and Town: A Guide to Its Social History* (Kent, OH: Kent State University Press, 1985), 67–68.

5. "Aluminum Skyscraper," *Popular Mechanics* 100, no. 6 (December 1953): 87.

6. Julian M. Allwood and Jonathan M. Cullen, *Sustainable Materials with Both Eyes Open* (Cambridge: UIT Cambridge, 2012), 101.

7. "Paul Louis Toussaint Héroult," in Lance Day and Ian McNeil, eds., *Biographical Dictionary of the History of Technology* (New York: Routledge, 2002), 589–590.

8. Ludovic Cailluet, "Selective Adaptation of American Management Models: The Long-Term Relationship of Pechiney with the United States," in Matthias Kipping and Ove Bjarnar, eds., *The Americanisation of European Business: The Marshall Plan and the Transfer of U.S. Management Models* (New York: Routledge, 1998), 190–207.

9. Ivan Amato, *Stuff: The Materials the World Is Made Of* (New York: Avon, 1997), 59.

10. Ibid., 60.

11. Tom D. Crouch, *Wings: A History of Aviation from Kites to the Space Age* (New York: Norton, 2003), 168.

12. Ibid., 170.

13. Ibid., 171.

14. Ibid., 171.

15. Ibid., 321–322.

16. Ibid., 323.

17. Ibid., 324.

18. Ibid., 330.

19. Ibid., 331.

20. Thomas P. Hughes, *Rescuing Prometheus: Four Monumental Projects That Changed Our World* (New York: Pantheon, 1998).

21. Mimi Sheller, *Aluminum Dreams: The Making of Light Modernity* (Cambridge, MA: MIT Press, 2014), 18.

22. Smith, *From Monopoly to Competition*, 215.

23. Harry H. Stein, "Fighting for Aluminum and for Itself: The Bonneville Power Administration, 1939–1949," *Pacific Northwest Quarterly* 99, no. 1 (Winter 2007–2008): 3–15; Eric Schatzberg, *Wings of Wood, Wings of Metal: Culture and Technical Choice in American Airplane Materials, 1914–1945* (Princeton, NJ: Princeton University Press, 1999), 192–195, 217–218, 220–221; Eric Schatzberg, "Symbolic Culture and Technological Change: The Cultural History of Aluminum as an Industrial Material," *Enterprise and Society* 4, no. 2 (June 2003): 226–271.

24. Felix Padel and Samarendra Das, *Out of This Earth: East India Adivasis and the Aluminum Cartel* (Delhi: Orient Blackswan, 2010), 72.

25. US Secretary of the Interior, *Annual Report of the Secretary of the Interior for the Fiscal Year Ended June 30, 1944* (Washington, DC: Government Printing Office, 1944), 5; Stein, "Fighting for Aluminum and for Itself," 13.

26. Crouch, *Wings*, 392–393.

27. Smith, *From Monopoly to Competition*, 234–242.

28. Charlotte Muller, "Aluminum and Power Control," *Journal of Land and Public Utility Economics* 21, no. 2 (May 1945): 108.

29. Ibid., 112.

30. Ibid., 113.

31. James E. Collier, "Aluminum Industry of Europe," *Economic Geography* 22, no. 2 (April 1946): 75.

32. Ibid., 75–108.

33. Smith, *From Monopoly to Competition*, 215.

34. Hans Otto Frøland, "Nazi Germany's Pursuit of Bauxite and Alumina," in Robin S. Gendron, Mats Ingulstad, and Espen Storli, eds., *Aluminum Ore: The Political Economy of the Global Bauxite Industry* (Vancouver: University of British Columbia Press, 2013), 79–106.

35. Sam H. Patterson et al., *World Bauxite Resources*, US Geological Survey Paper 1076-B (Washington, DC: Government Printing Office, 1986), B80–B84.

36. Edwin J. Mejia, "Aluminum in World Affairs," *Analysts Journal* 8, no. 5 (November 1952): 39–41.

37. Ronald Graham, *The Aluminum Industry and the Third World: Multinational Corporations and Underdevelopment* (London: Zed, 1982), 26–27; Smith, *From Monopoly to Competition*, 191.

38. US Geological Survey, "Aluminum Statistics," in T. D. Kelly and G. R. Matos, comps., *Historical Statistics for Mineral and Material Commodities in the United States: U.S. Geological Survey Data*, ser. 140. http://minerals.usgs.gov/minerals/pubs/historical-statistics/ds140-alumi.xlsx (accessed 21 December 2015).

39. Smith, *From Monopoly to Competition*, 215.

40. Ibid., 216.

41. Ibid., 216–217.

42. Ibid.

43. Ibid., 223.

44. Ibid., 234.

45. Ibid., 192.

46. Sheller, *Aluminum Dreams*, 49.

47. Ibid., 50.

48. Smith, *From Monopoly to Competition*, 192.

49. Crouch, *Wings*, 401.

50. Ibid.

51. Ibid., 402–403.

52. Harry N. Holmes, *Fifty Years of Industrial Aluminum* (Oberlin, OH: Oberlin College Bulletin, 1937); Alfred Cowles, *The True Story of Aluminum* (Chicago: Regnery, 1958); Jonathan M. Mitchell, *Aluminum: The First One Hundred Years* (Surrey, England: Fuel and Metallurgical Journals, 1986); Wendy Kaplan, *Designing Modernity: The Arts of Reform and Persuasion, 1885–1945* (New York: Thames and Hudson, 1995).

53. Quoted in Sheller, *Aluminum Dreams*, 61.

54. Annmarie Brennan, "Forecast," in Beatriz Colomina, Annemarie Brennan, and Jeannie Kim, eds., *Cold War Hothouses: Inventing Postwar Culture, from Cockpit to Playboy* (New York: Princeton Architectural Press, 2004), 60; Sheller, *Aluminum Dreams*, 70.

55. "General Notes," *Journal of the Royal Society of Arts* 95, no. 4735 (17 January 1947): 148–149; Clive Edwards, "Technology Transfer and the British Furniture Making Industry, 1945–1955," *Comparative Technology Transfer and Society* 2, no. 1 (April 2004): 71–98.

56. Hugh B. Johnston, "The Aluminum Industry and Design," *Industrial Design* 3 (August 1956): 50–63.

57. Frank Magee, "Alcoa and Design," in Samuel Fahnestock, ed., *Design Forecast 1* (Pittsburgh, PA: Alcoa, 1959), 4.

58. Dennis P. Doordan, "Promoting Aluminum: Designers and the American Aluminum Industry," *Design Issues* 9, no. 2 (Autumn 1993): 44–50.

59. Judith Dupré, *Skyscrapers: A History of the World's Most Extraordinary Buildings* (New York: Black Dog and Leventhal, 2013), 44.

60. "Architecture for the Future: GM Constructs a Versailles for Industry," *Life* 40, no. 21 (21 May 1956): 102–107; Eeva-Liisa Pelkonen and Donald Albrecht, eds., *Eero Saarinen: Shaping the Future* (New York: Finnish Cultural Institute, 2006), 1, 162–164.

61. US Geological Survey, "Aluminum Statistics"; Cowles, *The True Story of Aluminum*, 1.

62. Margaret B. Graham and Bettye H. Pruitt, *R&D for Industry: A Century of Technical Innovation at Alcoa* (New York: Cambridge University Press, 1990), 308–309.

63. US Congress, Office of Technology Assessment, *Nonferrous Metals: Industry Structure—Background Paper*, OTA-BP-E-62 (Washington, DC: Government Printing Office, 1990), 30.

64. US Geological Survey, "Aluminum Statistics."

65. Timothy J. LeCain, *Mass Destruction: The Men and Giant Mines That Wired America and Scarred the Planet* (New Brunswick, NJ: Rutgers University Press, 2009), 8.

66. Sheller, *Aluminum Dreams*, 16.

67. Ibid., 17.

68. Jennifer Gitlitz, *Trashed Cans: The Global Environmental Impacts of Aluminum Can Wasting in America* (Arlington, VA: Container Recycling Institute, 2002), 17.

69. Lloyd B. Coke, Colin C. Weir, and Vincent G. Hill, "Environmental Impact of Bauxite Mining and Processing in Jamaica," *Social and Economic Studies* 36, no. 1 (March 1987): 289–325, 327–330, 332–333; Helen McBain, "The Impact of the Bauxite-Alumina MNCs on Rural Jamaica: Constraints on Development of Small Farmers in Jamaica," *Social and Economic Studies* 36, no. 1 (March 1987): 137–170; Adun Ruud, "Can Transnational Aluminium Producers Be Ecologically Sustainable? A Case Study of Jamaica's Bauxite/Alumina Industry," *Business Strategy and the Environment* 3, no. 2 (Summer 1994): 82; Vidhya Das, "Mining Bauxite, Maiming People," *Economic and Political Weekly* 36, no. 28 (July 14–20, 2001): 2612–2614; Christopher T. Perry and Kevin G. Taylor, "Impacts of Bauxite Sediment Inputs on a Carbonate-Dominated Embayment, Discovery Bay, Jamaica," *Journal of Coastal Research* 20, no. 4 (Autumn 2004): 1070–1079.

70. Vincent Wright, Stephen Jones, and Felix Omoruyi, "Effect of Bauxite Mineralized Soil on Residual Metal Levels in Some Post Harvest Food Crops in Jamaica," *Bulletin of Environmental Contamination and Toxicology* 89, no. 4 (October 2012): 829.

71. M. C. Friesen et al., "Relationships between Alumina and Bauxite Dust Exposure and Cancer, Respiratory and Circulatory Disease," *Occupational and Environmental Medicine* 66, no. 9 (September 2009): 615–618.

72. Sheller, *Aluminum Dreams*, 19.

73. Matthew Evenden, "Aluminum, Commodity Chains, and the Environmental History of the Second World War," *Environmental History* 16, no. 1 (January 2011): 69–93.

74. Ibid., 70.

75. Gitlitz, *Trashed Cans*, 9.

76. László Rétvári, "Natural Resources and the Environmental Problems of Their Utilization in Hungary," *GeoJournal* 32, no. 4 (April 1994): 337–342; Zsuzsa Gille, *From the Cult of Waste to the Trash Heap of History: The Politics of Waste in Socialist and Postsocialist Hungary* (Bloomington: University of Indiana Press, 2007), 224n20.

77. Kemi Fuentes-George, "Neoliberalism, Environmental Justice, and the Convention on Biological Diversity: How Problematizing the Commodification of Nature Affects Regime Effectiveness," *Global Environmental Politics* 13, no. 4 (November 2013): 144–163.

CHAPTER TWO: **Designing Waste**

1. Chris Jordan, *Cans Seurat, 2007*, in *Running the Numbers: An American Self-Portrait*. http://www.chrisjordan.com/gallery/rtn/#cans-seurat (accessed 19 October 2013).

2. "Ann Wizer," in *CP Biennale 2006 Urban/Culture* (Jakarta: Kepustakaan Populer Gramedia, 2005), 231; Timo Rissanen and Holly McQuillan, *Zero Waste Fashion Design* (New York: Fairchild, 2016); Jared Hatch, "Arti Sandhu Walks Us through 'ZERØ Waste: Fashion Re-Patterned,'" *Chicago Racked*, 8 March 2011. http://chicago.racked.com /2011/3/8/7771593/curator-arti-sandhu-walks-us-through-zero-waste-fashion-repatterned; Chris Jordan, *Caps Seurat, 2011*, in *Running the Numbers: An American Self-Portrait*. http://www.chrisjordan.com/gallery/rtn/#caps-seurat (accessed 19 October 2013).

3. Thomas P. Hughes, *Rescuing Prometheus: Four Monumental Projects That Changed*

Our World (New York: Pantheon, 1998), 4, 105; Tom Crouch, *Wings: A History of Aviation from Kites to the Space Age* (New York: Norton, 2003), 516–517.

4. Lizabeth Cohen, *A Consumers' Republic: The Politics of Mass Consumption in Postwar America* (New York: Knopf, 2003), 8.

5. Karal Ann Marling, *As Seen on TV: The Visual Culture of Everyday Life in the 1950s* (Cambridge, MA: Harvard University Press, 1994), 243–283; Gail Cooper, "House and Home," in Carroll Pursell, ed., *A Companion to American Technology* (Malden, MA: Blackwell, 2005), 83–96; Ruth Oldenziel and Karin Zachman, "Kitchens as Technology and Politics: An Introduction," in Ruth Oldenziel and Karin Zachman, eds., *Cold War Kitchen: Americanization, Technology, and European Users* (Cambridge, MA: MIT Press, 2009), 1–32.

6. Eric Schatzberg, "Symbolic Culture and Technological Change: The Cultural History of Aluminum as an Industrial Material," *Enterprise and Society* 4, no. 2 (2003): 249.

7. Ibid., 252–253, 255.

8. John Lauber, "And It Never Needs Painting: The Development of Residential Aluminum Siding," *APT Bulletin* 31, nos. 2–3 (2000): 17–24.

9. "Zoners O.K. Building Prefabricated House," *Louisville Courier-Journal*, 23 April 1946; "Reynolds Must Move or Remove Model Home," *Louisville Courier-Journal*, 15 August 1946.

10. "Look . . . Aluminum Shingles, Aluminum Clapboard," advertisement in *Saturday Evening Post* 219 (February 1947): 76–77.

11. "Kaiserville, Pop. 40,000," *Architectural Forum* 78 (February 1943): 34: "More Kaiser," *Architectural Forum* 81 (September 1944): 77; "Homes by Kaiser," *Business Week*, 19 May 1945, 41.

12. "Kaiser Aluminum: A Major Supplier in Three Short Years," *Modern Metals* 5 (June 1949): 24; "Model Home Stimulates Sales," *American Builder* 71 (July 1949): 150; "Kaiser Homes: Building 8 Houses a Day," *American Builder* 71 (January 1949): 127.

13. "Alside's Vision Expansion," *Akron Beacon-Journal*, 14 October 1958.

14. Stuart W. Leslie, "The Strategy of Structure: Architectural and Managerial Style at Alcoa and Owens-Corning," *Enterprise and Society* 12, no. 4 (December 2011): 864; "Modernique in Pittsburgh," *Baltimore Sun*, 5 January 1954, 14.

15. Dennis P. Doordan, "Promoting Aluminum: Designers and the American Aluminum Industry," *Design Issues* 9, no. 2 (Autumn 1993): 47.

16. Ibid.

17. Quoted in Samuel Fahnestock, ed., *Design Forecast 1* (Pittsburgh, PA: Alcoa, 1959), 4.

18. John Peter, *Aluminum in Modern Architecture* (Louisville, KY: Reynolds Metal, 1956), 2:13; Doordan, "Promoting Aluminum," 47.

19. Fahnestock, *Design Forecast 1*, 4.

20. Ibid., 62–63.

21. Ibid.

22. Ibid.

23. Ibid.

24. Ibid., 62.

25. Samuel L. Fahnestock, ed., *Design Forecast 2* (Pittsburgh, PA: Alcoa, 1960), 4.

26. Ibid.

27. Ibid., 10.

28. Ibid.

29. Vance Packard, *The Waste Makers* (New York: Van Reese, 1960), 6.

30. Ibid., 43.

31. Ibid., 44.

32. Ibid., 116.

33. Ibid.

34. Craig Vogel, "Aluminum: A Competitive Material of Choice in the Design of New Products, 1950 to the Present," in Sarah C. Nichols et al., eds., *Aluminum by Design* (Pittsburgh, PA: Carnegie Museum of Art, 2000), 142.

35. Phil Patton, *Made in USA: The Secret History of the Things That Made America* (New York: Grove Weidenfeld, 1992), 99.

36. Vogel, "Aluminum," 143.

37. Ibid.

38. Ibid.

39. Ibid., 145.

40. Ibid.

41. John A. Kouwenhoven, *The Beer Can by the Highway: Essays on What's American About America* (Garden City, NY: Doubleday, 1961).

42. Peter Blake, *God's Own Junkyard: The Planned Deterioration of America's Landscape* (New York: Holt, Rinehart and Winston, 1964), 69.

43. Fahnestock, *Design Forecast 2*, 35.

44. Vogel, "Aluminum," 149.

45. *Modern Metals* 27 (1971): 88; US Patent S3453949 A, "Broiling Pan," filed 13 November 1967. http://www.google.com/patents/US3453949 (accessed 3 December 2014).

46. US Patent D187304 S, "Tray for Meats or Similar Article," granted 23 February 1960. http://www.google.com/patents/US3155303; US Patent 3113505 A, "Disposable Broiling Tray," filed 5 September 1961. http://www.google.com/patents/US3113505 (both accessed 3 December 2014).

47. Fahnestock, *Design Forecast 2*, 35.

48. Vogel, "Aluminum," 152.

49. Hyla M. Clark, *The Tin Can Book: The Can as Collectable Art, Advertising Art, and High Art* (New York: New American Library, 1977), 32, 124; Ann Vileisis, "Are Tomatoes Natural?," in Martin Reuss and Stephen H. Cutcliffe, eds., *The Illusory Boundary: Environment and Technology in History* (Charlottesville: University of Virginia Press, 2010), 211–248.

50. Robert Friedel, "American Bottles: The Road to No Return," *Environmental History* 19, no. 3 (July 2014): 509.

51. James Harvey Young, *Pure Food: Securing the Federal Food and Drugs Act of 1906* (Princeton, NJ: Princeton University Press, 1989), 35–39.

52. Vogel, "Aluminum," 153.

53. Ibid.

54. Thomas Hine, *The Total Package: The Evolution and Secret Meanings of Boxes, Bottles, Cans, and Tubes* (Boston: Little, Brown, 1995), 161.

55. Dan Baum, *Citizen Coors: An American Dynasty* (New York: William Morrow, 2002), 38.

56. Vogel, "Aluminum," 153.

57. Ibid., 155.

58. Ibid.

59. Henry Petroski, *The Evolution of Useful Things* (New York: Knopf, 1992), 199–203.

60. Vogel, "Aluminum," 154.

61. Clark, *The Tin Can Book*, 11, 33.

62. Vogel, "Aluminum," 155.

63. George L. Henderson and Mary Van Beck, "Mathematics Educators Must Help Face the Environmental Pollution Challenge," *Arithmetic Teacher* 17, no. 7 (November 1970): 559.

64. Bartow J. Elmore, *Citizen Coke: The Making of Coca-Cola Capitalism* (New York: Norton, 2015), 233–234; "Heads New Anti-Litter Group," *New York Times*, 14 October 1954, 31.

65. Elmore, *Citizen Coke*, 234; "Litter Increased in Crowded Cities," *New York Times*, 7 December 1954, 40.

66. Elmore, *Citizen Coke*, 234–235.

67. Ibid.

68. Austin's 1968 remarks quoted ibid., 235.

69. Paul Austin, "Environmental Renewal or Oblivion . . . Quo Vadis?," speech to the Georgia Bankers Association, Atlanta, 16 April 1970, reprinted in *Vital Speeches of the Day* 36, no. 15 (15 March 1979): 474; Elmore, *Citizen Coke*, 236.

70. William L. Rathje and Cullen Murphy, *Rubbish! The Archaeology of Garbage* (New York: HarperCollins, 1992), 25–26.

71. US Patent 4268531 A, filed 10 November 1976. https://www.google.com/patents/US4268531; US Patent 4236652 A, filed 20 March 1979. https://www.google.com/patents/US4236652 (both accessed 16 April 2015).

72. Rathje and Murphy, *Rubbish!* 25–26.

CHAPTER THREE: **A Recyclable Resource**

1. Michael Kimmelman, "A Grace Note for a Gritty Business," *New York Times*, 18 November 2013, C1.

2. George H. Manlove, "Junk Pile Transformed into Gold," *Iron Trade Review*, 9 May 1918, 1173–1176.

3. On American recycling programs, see Martin V. Melosi, *The Sanitary City: Urban Infrastructure in America from Colonial Times to the Present* (Baltimore: Johns Hopkins University Press, 2000), 413. On the reverse vending machine's history, see Finn Arne Jørgensen, *Making a Green Machine: The Infrastructure of Beverage Container Recycling* (New Brunswick, NJ: Rutgers University Press, 2011).

4. US Geological Survey, "Aluminum Statistics," in T. D. Kelly and G. R. Matos, comps., *Historical Statistics for Mineral and Material Commodities in the United States: U.S. Geological Survey Data*, ser. 140. http://minerals.usgs.gov/minerals/pubs/historical-statistics/ds140-alumi.xlsx (accessed 21 December 2015).

5. "21,000 Planes Bring $6,582,146 as Scrap," *New York Times*, 29 August 1946, 38.

6. Ibid.

7. Ibid.

8. Howard W. Rasher, ed., *The Nonferrous Scrap Metal Industry: Its Operations, Procedures, Techniques* (New York: National Association of Secondary Material Industries, 1967), 107.

9. Jørgensen, *Making a Green Machine*, 24.

10. Charles Lipsett, *Industrial Wastes and Salvage: Their Conservation and Utilization* (New York: Atlas, 1951), 62.

11. Ibid., 66.

12. Ibid., 62–63.

13. Ibid., 64–65.

14. Ibid., 66.

15. Ibid.

16. Ibid.

17. Ibid., 66–67.

18. Ibid., 67.

19. US Geological Survey, "Aluminum Statistics."

20. "'Profit' by Air Force on Scrap Is Disputed," *New York Times*, 29 September 1953, 42.

21. US Geological Survey, "Aluminum Statistics."

22. "Big Aluminum Smelter," *New York Times*, 28 November 1956, 70.

23. "Scrap Aluminum Gets New Life," *New York Times*, 7 April 1959, 45.

24. Ibid.

25. Brendan M. Jones, "Demand Is Rising for Scrap Metal," *New York Times*, 6 April 1959, 39, 43.

26. Kenneth S. Smith, "Aluminum Men Exude Optimism," *New York Times*, 8 January 1962, 106.

27. US Geological Survey, "Aluminum Statistics."

28. Margaret B. Graham and Bettye H. Pruitt, *R&D for Industry: A Century of Technical Innovation at Alcoa* (New York: Cambridge University Press, 1990), 308.

29. Ibid., 309.

30. Charles S. Rosenblum, "Aluminum," in Michael Suisman and Howard W. Rasher, eds., *Nonferrous Scrap Metal Guidebook: Origin, Preparation, Usage, and Related Subjects* (New York: National Association of Secondary Material Industries, 1960), 9.

31. Ibid.

32. Ibid.

33. Charles Lipsett, *Industrial Wastes and Salvage: Their Conservation and Utilization*, 2nd ed. (New York: Atlas, 1963), 106.

34. Ibid., 108.

35. Ibid., 106.

36. Ibid.

37. Ibid.

38. Ibid.

39. Ibid., 106–107.

40. Ibid.

41. Historically, most recycled aluminum has been used in casting alloys. Julian M. Allwood and Jonathan M. Cullen, *Sustainable Materials with Both Eyes Open* (Cambridge: UIT Cambridge, 2012), 53.

42. Lipsett, *Industrial Wastes and Salvage* (1963), 112.

43. Ibid.

44. Rasher, *Nonferrous Scrap Metal Industry*, 28.

45. Ibid.

46. Ibid.

47. Ibid., 28–29.

48. Ibid., 29.

49. Ibid.

50. Ibid.

51. US Geological Survey, "Aluminum Statistics."

52. Paul Fine, Howard W. Rasher, and Si Wakesberg, eds., *Operations in the Nonferrous Scrap Metal Industry Today* (New York: National Association of Secondary Material Industries, 1973), 25.

53. Ibid., 27–28.

54. Peter Blake, *God's Own Junkyard: The Planned Deterioration of America's Landscape* (New York: Holt, Rinehart and Winston, 1964), 10; President Lyndon B. Johnson, "Remarks at the Signing of the Highway Beautification Act of 1965, October 22, 1965," Lyndon B. Johnson Library and Museum, Austin, TX (National Archives and Records Administration). http://www.lbjlib.utexas.edu/johnson/archives.hom/speeches.hom/651022.asp; "Highway Beautification Act: It's [sic] Effect on Scrap Industry," *Scrap Age* 22, no. 11 (1965): 16–17; Carl A. Zimring, " 'Neon, Junk, and Ruined Landscape': Competing Visions of America's Roadsides and the Highway Beautification Act of 1965," in Christof Mauch and Thomas Zeller, eds., *The World beyond the Windshield: Roads and Landscapes in the United States and Europe* (Athens: Ohio University Press, 2008), 94–107.

55. Blake, *God's Own Junkyard*, 69.

56. Samuel P. Hays, *Beauty, Health, and Permanence: Environmental Politics in the United States, 1955–1985* (New York: Cambridge University Press, 1987), 53, 80–83, 171–206; Joel Tarr, *The Search for the Ultimate Sink: Urban Pollution in Historical Perspective* (Akron, OH: University of Akron Press, 1996), 349.

57. "Heads New Anti-litter Group," *New York Times*, 14 October 1954, 31; Bartow J. Elmore, *Citizen Coke: The Making of Coca-Cola Capitalism* (New York: Norton, 2015), 233.

58. Bartow J. Elmore, "The American Beverage Industry and the Development of Curbside Recycling Programs, 1950–2000," *Business History Review* 86 (2012): 477–501.

59. Gladwin Hill, "Half-Cent Bounty on Old Cans Puts a Small Dent in Mountains of Trash," *New York Times*, 19 August 1969, 23.

60. Samantha MacBride, *Recycling Reconsidered: The Present Failure and Future Promise of Environmental Action in the United States* (Cambridge, MA: MIT Press, 2011), 57.

61. Battelle Memorial Institute, *A Study to Identify Opportunities for Increased Solid Waste Utilization* (Washington, DC: US Environmental Protection Agency, 1972), 2:xiv.

62. Ibid., 2.

63. Ibid., 60.

64. Ibid., 58.

65. Graham and Pruitt, *R&D for Industry*, 290.

66. Ibid., 419–420.

67. Ibid., 425.

68. Ibid., 432.

69. Ibid., 308.

70. National Association of Recycling Industries, *Recycling Aluminum* (New York: National Association of Recycling Industries, 1981), 2.

71. Ibid., 7.

72. Ibid., 12–14.

73. Ibid., 16.

74. Ibid.

75. US Geological Survey, "Aluminum Statistics."

76. Henry Petroski, *The Evolution of Useful Things* (New York: Knopf, 1992), 199–203.

77. National Institute for Occupational Safety and Health, *NIOSH Pocket Guide to Chemical Hazards.* http://www.cdc.gov/niosh/npg/npgd0023.html (accessed 27 February 2015).

78. William J. Harnisch, "Air Quality Control in the Metallurgical Industry," *Natural Resources Lawyer* 3, no. 1 (January 1970): 131–140.

79. "Chloride-Free Processing of Aluminum Scrap," *Aluminium Today* 8, no. 3 (June–July 1996): 18.

80. Bernadette Vielhaber, "A Successful Approach to MACT Compliance," *Foundry Management and Technology* 132, no. 11 (November 2004): 18–19; Lee Wei-Shan et al., "Emissions of Polychlorinated Dibenzo-p-Dioxins and Dibenzofurans from Stack Gases of Electric Arc Furnaces and Secondary Aluminum Smelters," *Journal of the Air and Waste Management Association* 55, no. 2 (February 2005): 219–226; Paul Schaffer, "Two Smelters Fined for High Dioxin Emissions," *American Metal Market* 113, no. 14-1 (11 April 2005): 7.

81. A. V. Bridgwater and C. J. Mumford, *Waste Recycling and Pollution Control Handbook* (New York: Van Nostrand Reinhold, 1979), 417.

82. James F. King, *International Scrap and Recycling Industry Handbook* (Abingdon, England: Woodhead, 2001): 15–16.

83. Fine, Rasher, and Wakesberg, *Operations in the Nonferrous Scrap Metal Industry*, 31.

84. Ibid.

85. Data from Right to Know Network, Toxics Release Inventory database. http://www.rtknet.org/db/tri/tri.php?database. Search term "331314: Secondary Smelting and Alloying of Aluminum"; search conducted 4 May 2015.

86. Timothy Erdman, "Secondary Market Scraps the Old for the New," *Metal Producing* 36, no. 9 (September 1998): N2-7; Peng Li et al., "Recycling of Aluminum Salt Cake: Utilization of Evolved Ammonia," *Metallurgical and Materials Transactions* 44, no. 1 (February 2013): 16–19.

87. Allwood and Cullen, *Sustainable Materials*, 111.

88. US Geological Survey, "Aluminum Statistics."

89. Melosi, *The Sanitary City*, 413.

90. European Aluminium Association and Organisation of European Aluminium Refiners and Remelters, *Aluminium Recycling in Europe: The Road to High Quality Products* (Meckenheim, Germany: EAA/OEARR Recycling Division, 2006), 8.

91. European Aluminium Association, "Environmental Profile Report for the European Aluminium Industry," April 2008, 49–50. http://european-aluminium.eu/media/1329/environmental-profile-report-for-the-european-aluminium-industry.pdf.

92. Novelis, "Our History." http://novelis.com/about-us/#1445023525125-635e2312-1c53 (accessed 14 December 2015).

93. Joe Trini, "Novelis Reports Record Can Recycling in 2006," *Waste News* 12, no. 25 (April 16, 2007): 11.

94. "Aluminum Recycling at Novelis." http://novelis.com/sustainability/novelis-recycling-capabilities (accessed 14 December 2015).

95. Ibid.

96. Marc Gunther, "Why Are Major Beverage Companies Refusing to Use a 90 Percent Recycled Can?," *Guardian*, 30 October 2014. http://www.theguardian.com/sustainable-business/2014/oct/30/recycled-aluminum-novelis-ford-cocacola-pepsi-miller-budweiser-beer.

97. Discussions of European Union regulations on individual and industrial waste and recycling practices include Allen R. Bailey and Melinda C. Bailey, *The EU Directive Handbook: Understanding the European Union Compliance Process and What It Means to You* (Boca Raton, FL: St. Lucie Press, 1997), 59–61; Geert van Calster, "Regulatory Instruments: Sustainable Materials Management, Recycling, and the Law," in Ernst Worrell and Markus A. Reuter, eds., *Handbook of Recycling: State-of-the-Art for Practitioners, Analysts, and Scientists* (Waltham, MA: Elsevier, 2014), 527–536.

98. K. J. Martchek, "Modelling More Sustainable Aluminum," *International Journal Life Cycle Assessment* 11, no. 1 (2006): 34–37; Wei-Qiang Chen, "Recycling Rates of Aluminum in the United States," *Journal of Industrial Ecology* 17, no. 6 (December 2013): 926–938.

CHAPTER FOUR: **Metal in Motion**

1. "The Future of Tough: New F-150," Ford Motor Company. http://www.ford.com/trucks/f150/?fmccmp=lp-trucks-top-hp-f-150 (accessed 2 June 2015).

2. Chaz Miller, "Profiles in Garbage: Aluminum Packaging," *Waste360* 45, no. 7 (1 September 2014): 28.

3. John David Anderson, *The Airplane: A History of Its Technology* (Reston, VA: American Institute of Aeronautics and Astronautics, 2002), 260–263.

4. Edward W. Constant II, *The Origins of the Turbojet Revolution* (Baltimore: Johns Hopkins University Press, 1980).

5. Ibid.

6. René J. Francillon, *Boeing 707: Pioneer Jetliner* (Osceola, WI: MBI Publishing, 1999), 24; Alten F. Grandt Jr., *Fundamentals of Structural Integrity: Damage Tolerant Design and Nondestructive Evaluation* (Hoboken, NJ: Wiley, 2004), 153.

7. "Boeing 707," *Aviation History Online Museum*. http://www.aviation-history.com/boeing/707.html (accessed 12 March 2015).

8. http://active.boeing.com/commercial/orders/displaystandardreport.cfm?cboCurrentModel=707&optReportType=AllModels&cboAllModel=707&ViewReportF=View+Report (accessed 17 June 2015).

9. http://active.boeing.com/commercial/orders/displaystandardreport.cfm?cboCur

rentModel=747&optReportType=AllModels&cboAllModel=747&ViewReportF=View+ Report (accessed 17 June 2015).

10. J. Randoph Kissell and Robert L. Perry, *Aluminum Structures: A Guide to Their Specifications and Design*, 2nd ed. (New York: Wiley, 2002), 24. Spacecraft and some recent airplanes, including the Boeing 787 Dreamliner, employ other materials in their bodies, including titanium alloys and carbon-fiber composites.

11. Gordon Baxter, "Banking on a Taildragger," *Flying Magazine* 106, no. 5 (May 1980): 62–63, 66.

12. David Hounshell, *From the American System to Mass Production, 1800–1932: The Development of Manufacturing Technology in the United States* (Baltimore: Johns Hopkins University Press, 1984), 190.

13. W. Bernard Carlson, *Innovation as a Social Process: Elihu Thomson and the Rise of General Electric, 1870–1900* (New York: Cambridge University Press, 1991), 332; Tony Hadland and Hans-Erhard Lessing with Nick Clayton and Gary W. Sanderson, *Bicycle Design: An Illustrated History* (Cambridge, MA: MIT Press, 2014), 311; David V. Herlihy, *Bicycle: The History* (New Haven, CT: Yale University Press, 2004), 7, 289.

14. Hadland and Lessing, *Bicycle Design*, 395.

15. Herlihy, *Bicycle*, 327.

16. Hadland and Lessing, *Bicycle Design*, 180.

17. Ibid., 395.

18. Ibid., 167.

19. Ibid.

20. Ibid., 396.

21. "German Maker of Kettcar Files for Insolvency," *Reuters*, 3 June 2015. http://www .reuters.com/article/2015/06/03/germany-kettler-idUSL5N0YP1ZY20150603.

22. Hadland and Lessing, *Bicycle Design*, 393.

23. Herlihy, *Bicycle*, 368.

24. Hadland and Lessing, *Bicycle Design*, 393.

25. Herlihy, *Bicycle*, 368.

26. Classified ad 13 (untitled), *Chicago Tribune*, 13 July 1989, NB12.

27. "Vintage Cannondale," discussion thread, *Adventure Cycling Association Forum*, 1 August 2010. http://forums.adventurecycling.org/index.php?topic=7544.0.

28. Hadland and Lessing, *Bicycle Design*, 395–396. The Aluetta bicycle was patented in Italy in 1986 (patent 19748A).

29. Ibid., 396.

30. Mike Yozell, "First Look: Cannondale CAAD12," *Bicycling*, 9 July 2015. http:// www.bicycling.com/bikes-gear/reviews/first-look-cannondale-caad12.

31. Geoffrey Davies, *Materials for Automobile Bodies*, 2nd ed. (Waltham, MA: Butterworth-Heinemann, 2012), 48.

32. Nick Baldwin and G. N. Georgano, *The World Guide to Automobile Manufacturers* (New York: Facts on File, 1987), 209.

33. Robert Cass, "The Automobile Industry," *Analysts Journal* 9, no. 3 (June 1953): 69.

34. Ibid.

35. Tom McCarthy, *Auto-Mania: Cars, Consumers, and the Environment* (New Haven, CT: Yale University Press, 2007), 125–128.

36. Reynolds Metals Company, *Aluminum in Automobiles* (Richmond, VA: Reynolds Metals, 1959), 7.

37. Ibid., 9.

38. Ibid., 14.

39. Ibid.

40. Ibid., 10.

41. Ibid.

42. Ibid., 11.

43. Friedrich Ostermann, "Aluminum Materials for Vehicle Construction," in his *Aluminum Materials Technology for Automobile Construction*, trans. Pam Chatterly (London: Mechanical Engineering Publications, 1993), 1.

44. Reynolds Metals Company, *Aluminum in Automobiles*, 10.

45. René Bellu, "Toutes les Voitures Françaises 1955 (Salon Paris, Octobre 1954)," in Bellu, ed., *Automobilia* (Paris: Histoire et Collections, 1999), 10, 46, 47, 51.

46. Hermann E. Burst and Erich W. Strehler, *The Use of Aluminum in the Porsche 928* (Warrendale, PA: Society of Automotive Engineers, 1978); K. M. Loasby, *The Use and Manipulation of Aluminum in Aston Martin and Lagonda Cars* (Warrendale, PA: Society of Automotive Engineers, 1978); Tim Cottingham, "DB2/4," *Aston Martins*. http://astonmartins.com/car/db24 (accessed 10 December 2014).

47. "As Featured in 'The Birds': Aston Martin DB2/4 Drophead Coupé," *Classic Driver*, 29 April 2013. https://www.classicdriver.com/en/article/cars/featured-birds-aston-martin-db24-drophead-coupe?language=en-US; Phil Patton, "At Auction, Masterpieces of the Drivable Kind," *New York Times*, 17 November 2013, AU7.

48. Rob Leicester Wagner, *Style and Speed: The World's Greatest Sports Cars* (New York: MetroBooks, 1998), 57; Winston Goodfellow, *Ferrari Hypercars: The Inside Story of Maranello's Fastest, Rarest Cars* (Minneapolis, MN: Motorbooks, 2014), 163.

49. John Lamm, *Ferrari: Stories from Those Who Lived the Legend* (St. Paul, MN: Motorbooks, 2007), 98–99; Douglas Martin, "Sergio Scaglietti, 91, Sculptor of Sleekly Tailored Ferraris," *New York Times*, 27 November 2011, A30.

50. Lamm, *Ferrari*, 99.

51. John Lamm, *Supercars* (Osceola, WI: MBI Publishing, 2001), 56; Leonardo Acerbi, *Ferrari: A Complete Guide to All Models* (St. Paul, MN: Motorbooks, 2005), 338–339.

52. Wagner, *Style and Speed*, 57; Goodfellow, *Ferrari Hypercars*, 163.

53. Doug Mitchel and Tom Collins, *Supercars Field Guide* (Iola, WI: KP Books, 2006), 287.

54. Burst and Strehler, *The Use of Aluminum in the Porsche 928*, 1–2.

55. Ibid., 2.

56. Sohan L. Chawla and Rajeshwar K. Gupta, *Materials Selection for Corrosion Control* (Materials Park, OH: ASM International, 1993), 192.

57. Ostermann, "Aluminum Materials for Vehicle Construction," 6.

58. "Light-Artikel: Leichtbau-Material Aluminium," *Auto Motor und Sport* 24 (15 November 1991): 100–103.

59. Ostermann, "Aluminum Materials for Vehicle Construction," 11.

60. Dan McCosh, "Material Wealth," *Popular Science* 237, no. 2 (August 1990): 83.

61. Davies, *Materials for Automobile Bodies*, 48.

62. Michael Lamm, "Original Acura NS-X Buyer's Guide: A Mid-Engined Supercar for the Price of a New Accord," *Car and Driver*, 11 December 2012. http://blog.carand driver.com/original-acura-nsx-buyers-guide-a-mid-engined-supercar-for-the-price-of-a -new-accord.

63. Alcoa Automotive Structures, *Life Cycle Energy Savings through the Use of Light-weight Aluminum Body Structures* (Pittsburgh, PA: Alcoa, 1994), 1.

64. Ibid.

65. Michael Thompson, *Rubbish Theory: The Creation and Destruction of Value* (New York: Oxford University Press, 1979), 7–17.

66. Adam J. Gesing and Aron Rosenfeld, "Composition Based Sorting of Aluminum Scrap from Aluminum Intensive Vehicles," in *Developments in Aluminum Use for Vehicle Design* (Warrendale, PA: Society of Automotive Engineers, 1996), 29.

67. Ibid., 36.

68. Ibid., 30.

69. Ibid.

70. "Spotlight: Reinventing the Ford F-150." http://corporate.ford.com/micro sites/sustainability-report-2014-15/environment-spotlight-f150.html (accessed 22 May 2015).

71. Gary Witzenburg, "Pete Reyes, Chief Engineer: 2015 Ford F-150 Interview," *Truck Trend*, 19 September 2014. http://www.trucktrend.com/features/1411-pete-reyes-chief -engineer-2015-ford-f-150-interview.

72. Ibid.

73. "Ford's Aluminum F-150 Almost Ready for Prime Time," *Chicago Sun-Times*, 11 November 2014. http://chicago.suntimes.com/?p=146403.

74. Ibid.

75. Alissa Priddle, "CEO Phil Martens Leaves Aluminum Supplier Novelis," *Detroit Free Press*, 20 April 2015. http://www.freep.com/story/money/cars/2015/04/20/phil-mar tens-out-ceo-novelis-aluminum-ford-f150/26086977.

76. "Novelis Commissions $48 Million Automotive Closed-Loop Recycling Invest-ment in Oswego Plant," *PRNewswire*, 26 January 2015. http://www.prnewswire.com/news -releases/novelis-commissions-48-million-automotive-closed-loop-recycling-investment -in-oswego-plant-300025082.html.

77. Joann Muller, "How Novelis Is Lightening Ford's Load on the New F-150 Pickup," *Forbes*, 5 November 2014. http://www.forbes.com/sites/joannmuller/2014/11/05/novelis -is-lightening-fords-load-on-new-f-150.

78. John Gardner, "Building a Circular Economy: How Ford, Novelis Created a Truly Closed Loop for Automotive Aluminum," *Sustainable Brands*, 28 May 2015. http://www .sustainablebrands.com/news_and_views/next_economy/john_gardner/building_circu lar_economy_how_ford_novelis_created_truly_cl.

79. Ibid.

80. Andrew Wendler, "Aluminati: Official 2015 Ford F-150 EPA Fuel-Economy Num-bers Released," *Car and Driver*, 21 November 2014. http://blog.caranddriver.com/alumi nati-official-2015-ford-f-150-epa-fuel-economy-numbers-released.

81. "2015 Ford F-150 Pickup Truck." http://www.ford.com/trucks/f150 (accessed 2 June 2015).

82. Ben Geier, "Ford's F-150 Truck Had a Good First Month," *Fortune*, 3 February 2015. http://fortune.com/2015/02/03/fords-f-150-truck-had-a-good-first-month.

83. Ostermann, "Aluminum Materials for Vehicle Construction," 11; Brent Snavely, "First 2017 Acura NSX Supercar Comes Off Line in Ohio," *Detroit Free Press*, 24 May 2016. http://www.freep.com/story/money/cars/2016/05/24/production-2017-acura-nsx-super car-begins-ohio/84829634.

84. David N. Lucsko, *Junkyards, Gearheads, and Rust: Salvaging the Automotive Past* (Baltimore: Johns Hopkins University Press, 2016).

85. David N. Lucsko, "Of Clunkers and Camaros: Accelerated Vehicle Retirement Programs and the Automobile Enthusiast, 1990–2009," *Technology and Culture* 55, no. 2 (April 2014): 407.

CHAPTER FIVE: **Covetable Aluminum Furniture**

1. "Eames Chair," *Design within Reach*. http://dwr.com (accessed 2 June 2015).

2. "Top Ten Winners: Herman Miller Greenhouse Factory and Offices," American Institute of Architects. http://www.aiatopten.org/node/231 (accessed 17 March 2015).

3. Peter Sulzer, *Jean Prouvé: Oeuvres Complètes/Complete Works*, vol. 2: *1934–1944* (Basel: Birkhäuser Architecture, 1999), 139; Jean Prouvé, *La Maison Tropicale* (Paris: Centre Pompidou, 2009).

4. H. Ward Jandl, "With Heritage So Shiny: America's First All-Aluminum House," *APT Bulletin* 23, no. 2 (1991): 38–43.

5. "Aluminium in the Construction of Furniture," *Metal Industry*, 1 May 1931, 460.

6. John Peter, *Aluminum in Modern Architecture* (Louisville, KY: Reynolds Metals, 1956), 2:248.

7. Edwards, "Aluminium Furniture," 214.

8. Ibid., 215; "The Light Metal Home," *Light Metals*, October 1946, 439–561.

9. Edwards, "Aluminium Furniture," 217.

10. Ibid., 220; "The Booming Furniture Market," *Modern Metals* 6 (June 1950): 21.

11. Hugh De Pree, *Business as Unusual: The People and Principles at Herman Miller* (Zeeland, MI: Herman Miller, 1986), 17–18.

12. Ibid., 22.

13. Ibid., 24–25.

14. Interview of George Nelson by Mickey Friedman, 15 October 1974, Herman Miller, Zeeland, MI. Herman Miller Collection, the Henry Ford, Dearborn, MI. Acc 89.177, box 32, folder: Oral Transcript George Nelson.

15. De Pree, *Business as Unusual*, 53.

16. Marilyn Neuhart with John Neuhart, *The Story of Eames Furniture*, bk. 2: *The Herman Miller Age* (Berlin: Gestalten, 2010), 689.

17. Ibid., 698.

18. John R. Berry, *Herman Miller: The Purpose of Design*, rev. ed. (New York: Rizzoli International, 2009), 3.

19. Oral history interview with Ray Eames, 28 July–20 August 1980, Archives of American Art, Smithsonian Institution, Washington, DC.

20. Ralph Caplan, *The Design of Herman Miller* (New York: Whitney Library of Design, 1976), 43.

21. Leslie Piña, *Classic Herman Miller* (Atglen, PA: Schiffer, 1998), 7.

22. John Neuhart, Marilyn Neuhart, and Ray Eames, *Eames Design: The Work of the Office of Charles and Ray Eames* (New York: Abrams, 1989), 97.

23. Ibid.

24. Ibid., 98–99.

25. Ibid., 100–101.

26. Ibid., 207.

27. Marilyn J. Neuhart interviews with Don Albinson in El Segundo, CA, and Coopersburg, PA, 1995–2000, quoted in Neuhart with Neuhart, *The Story of Eames Furniture*, 662.

28. Ibid.

29. De Pree, *Business as Unusual*, 48.

30. Neuhart with Neuhart, *The Story of Eames Furniture*, 661.

31. Neuhart, Neuhart, and Eames, *Eames Design*, 227.

32. Neuhart interviews with Albinson, quoted in Neuhart with Neuhart, *The Story of Eames Furniture*, 662.

33. Neuhart with Neuhart, *The Story of Eames Furniture*, 665–666.

34. Interview of Charles and Ray Eames by Mickey Friedman, 6 November 1974, Walker Art Center, Minneapolis, MN. Herman Miller Collection, the Henry Ford, Dearborn, MI. Acc 89.177, box 32, folder: Oral Transcript Charles & Ray Eames.

35. Neuhart interviews with Albinson, quoted in Neuhart with Neuhart, *The Story of Eames Furniture*, 662.

36. Ibid.

37. Neuhart with Neuhart, *The Story of Eames Furniture*, 663.

38. Ibid.

39. Ibid., 663–664.

40. Ibid., 664.

41. Neuhart interviews with Albinson, quoted in Neuhart with Neuhart, *The Story of Eames Furniture*, 664.

42. Ibid.

43. Ibid., 664–665.

44. Charles Eames, "3 Chairs / 3 Records of the Design Process," *Interiors*, April 1958, quoted in Neuhart with Neuhart, *The Story of Eames Furniture*, 674.

45. Ibid.

46. Neuhart interviews with Albinson, quoted in Neuhart with Neuhart, *The Story of Eames Furniture*, 665–666.

47. Ibid., 666.

48. Ibid.

49. Ibid., 665–666.

50. Ibid., 666–667.

51. Ibid., 667–668.

52. Ibid., 668.

53. Ibid., 673.

54. Ibid., 673–674.

55. Ibid., 674.

56. Ibid.

57. Neuhart, Neuhart, and Eames, *Eames Design*, 227–228.

58. Ibid.

59. Bob Staples interview, Eames Office Oral History Project, Washington, DC, 13 March 13 1992; Eames Demetrios, *An Eames Primer* (New York: Universe Publishing, 2001), 176.

60. Neuhart, Neuhart, and Eames, *Eames Design*, 275.

61. Neuhart with Neuhart, *The Story of Eames Furniture*, 731.

62. Ibid., 735.

63. Ibid., 731.

64. Neuhart, Neuhart, and Eames, *Eames Design*, 275.

65. Ibid., 280.

66. Ibid., 281.

67. Craig Vogel, "Aluminum: A Competitive Material of Choice in the Design of New Products, 1950 to the Present," in Sarah C. Nichols et al., eds., *Aluminum by Design* (Pittsburgh, PA: Carnegie Museum of Art, 2000), 143.

68. Ibid.

69. Ibid., 147.

70. Ibid.

71. Ibid., 148.

72. Samuel Fahnestock, ed., *Design Forecast 1* (Pittsburgh, PA: Alcoa, 1959), 17.

73. Ibid.

74. Neuhart, Neuhart, and Eames, *Eames Design*, 243.

75. Ibid., 247.

76. Neuhart with Neuhart, *The Story of Eames Furniture*, 702.

77. Interview of Charles and Ray Eames by Mickey Friedman.

78. Neuhart, Neuhart, and Eames, *Eames Design*, 249.

79. Interview of Charles and Ray Eames by Mickey Friedman.

80. Interview of George Nelson by Mickey Friedman.

81. Interview of Charles and Ray Eames by Mickey Friedman.

82. Interview of George Nelson by Mickey Friedman.

83. Neuhart, Neuhart, and Eames, *Eames Design*, 283.

84. Ibid., 327.

85. Ibid., 343.

86. Ibid., 354.

87. Ibid., 371.

88. Ibid., 451.

89. Piña, *Classic Herman Miller*; Sarah C. Nichols et al., eds., *Aluminum by Design* (Pittsburgh, PA: Carnegie Museum of Art, 2000); Berry, *Herman Miller*; Victoria Ballard Bell and Patrick Rand, *Materials for Design* (New York: Princeton Architectural Press, 2006).

90. Interview of Charles and Ray Eames by Mickey Friedman.

91. Berry, *Herman Miller*, 24.

92. Marcus Fairs, *Green Design: Creative Sustainable Designs for the Twenty-First Cen-

tury (Berkeley, CA: North Atlantic Books, 2009), 71–75; "Charles Pollock, Designer of Popular Office Chair, Dies at 83," *New York Times*, 25 August 2013, A20.

93. Adrienne Breaux, "Interview: Gregg Buchbinder of Emeco," *2Modern*, 18 October 2011. http://blog.2modern.com/2011/10/interview-gregg-buchbinder-of-emeco.html.

94. De Pree, *Business as Unusual*, 160.

95. Ibid.

96. Interview of George Nelson by Mickey Friedman.

97. 1987 memorial booklet, University of Michigan, Ann Arbor. Herman Miller Collection, the Henry Ford, Dearborn, MI. Acc 89.177, box 31, folder: Designer: George Nelson.

98. Ibid.

99. Ibid.

100. Ibid.

101. Lloyd Alter, "Herman Miller's GreenHouse Factory Generates 15 Pounds of Landfill Waste Per Month," *Treehugger*, 17 February 2011. http://www.treehugger.com /sustainable-product-design/herman-millers-greenhouse-factory-generates-15-pounds -of-landfill-waste-per-month.html.

102. John Gertsatkis, Nicola Morelli, and Chris Ryan, "Industrial Ecology and Extended Producer Responsibility," in Robert U. Ayres and Leslie W. Ayres, eds., *A Handbook of Industrial Ecology* (Northampton, MA: Elgar, 2002), 525.

103. Ibid.

104. Mark Rossi et al., "Design for the Next Generation: Incorporating Cradle-to-Cradle Design into Herman Miller Products," *Journal of Industrial Ecology* 10, no. 4 (2006): 193–210.

105. Ibid., 195

106. Ibid., 204.

107. Herman Miller, *Our Journey toward a Better World around You* (Zeeland, MI: Herman Miller, 2012), 22.

108. Alter, "Herman Miller's GreenHouse Factory."

109. Samantha MacBride, *Recycling Reconsidered: The Present Failure and Future Promise of Environmental Action in the United States* (Cambridge, MA: MIT Press, 2011), 188.

110. "Norman Foster, Architect (1935–)," *Design Museum*. http://design.designmu seum.org/design/norman-foster (accessed 13 November 2013).

111. "Sir Norman Foster: The Master Builder," *Guardian*, 1 January 1999. http://www .theguardian.com/books/1999/jan/02/books.guardianreview10.

112. "Norman Foster, Architect."

113. Fairs, *Green Design*, 76.

114. Sarah C. Nichols, "Highlights," in Nichols et al., eds., *Aluminum by Design* (Pittsburgh, PA: Carnegie Museum of Art, 2000), 140–165, quote on 258.

115. Ibid.

116. "Hudson Task Chair, Polished," *Design Within Reach*. http://www.dwr.com/prod uct/hudson-task-chair-polished.do?sortby=ourPicks (accessed 2 June 2015).

117. Nichols, "Highlights," 270.

CHAPTER SIX: **Guitar Sustain**

1. "Auction of Grateful Dead Items Nets $1.1 million," *Marin Independent Journal*, 10 May 2007; Douglas Martin, "Travis Bean, 63, Aluminum Guitar Maker," *New York Times*, 16 July 2011, D8.

2. George Gruhn, "Rickenbacker A22 Frying Pan," *Vintage Guitar*, March 2004. http://www.vintageguitar.com/1969/rickenbacker-a22.

3. Ibid.; Michael John Simmons, "Catch of the Day: 1935 Gibson E-150," *Fretboard Journal*. http://www.fretboardjournal.com/blog/catch-day-1935-gibson-e-150 (accessed 8 December 2014).

4. Bill C. Malone, *Country Music USA* (Austin: University of Texas Press, 1985), 157–158, 203; Tom Wheeler, *American Guitars: An Illustrated History* (New York: HarperPerennial, 1992), 333–334; Jean A. Boyd, *The Jazz of the Southwest: An Oral History of Western Swing* (Austin: University of Texas Press, 1998), 114–116.

5. Michael Wright, "The Weird, Wacky World of Wandré Guitars," *Vintage Guitar* 28, no. 10 (August 2014): 138.

6. Ibid.

7. Marc O'Hara, "Wandré Guitars—The First Metal Necks?," *Unique Guitar*, 16 October 2009. http://uniqueguitar.blogspot.com/2009/10/wandre-guitars-first-metal-necks.html.

8. Frank Goodman, "Interview with Buddy and Julie Miller," *Pure Music* (October 2001). http://www.puremusic.com/pdf/buddyandjulie.pdf.

9. Ibid.

10. Ward Meeker, "The First 'Aluminum Neck Guitar' Built by Glen Burke," *Guitar Archaeology*, 2 May 2012. https://guitararchaeology.wordpress.com/2012/05/02/the-first-aluminum-neck-guitar-built-by-glen-burke-2; post on "Questions for the Aluminati" thread, *Electrical Audio*, 24 February 2013. http://www.electricalaudio.com/phpBB/viewtopic.php?f=4&t=61598&p=1784102#p1619164.

11. Art Thompson, "Messenger Guitars," *Guitar Player*, 20 April 2006. http://www.guitarplayer.com/miscellaneous/1139/messenger-guitars/20192; "1967 Musicraft Messenger Guitar," *Premier Guitar*, 11 February 2010. http://www.premierguitar.com/articles/1967_Musicraft_Messenger_Guitar.

12. Michael Wright, *Guitar Stories*, vol. 2: *The Histories of Cool Guitars* (Bismarck, ND: Vintage Guitar Books, 2000), 241.

13. Ibid.

14. Ibid.

15. Ibid., 242.

16. Ibid., 243.

17. Ibid., 243–244.

18. Ibid., 244.

19. Ibid.

20. Don Menn, "What's in a Name: Travis Bean," *Guitar Player*, 13 April 1979, 146.

21. Ibid.

22. Ibid., 148.

23. Art Thompson, "Artifacts: Travis Bean," *Guitar Player*, December 2005, 140.

24. Menn, "What's in a Name," 148.

25. Ibid.

26. Ibid., 148.

27. The 1978 interview is recounted in Thompson, "Artifacts: Travis Bean," 140.

28. Ibid.

29. Menn, "What's in a Name," 150.

30. Ibid., 152.

31. "Travis Bean," *Times of London*, 9 September 2011, 63.

32. Willie G. Moseley, "Travis Bean Interview: Metal Machine Music—The Next Phase," *Vintage Guitar Magazine*, January 1999. http://www.vintageguitar.com/1830/travis-bean-interview-2.

33. Eric Gaer, "Towards a Brave New Present," *Creem* 7 (January 1976): 56.

34. Ibid.

35. Ibid.

36. Ibid.

37. Allen Hester, "The Lean, Mean Travis Bean," *Creem* 8 (November 1976): 50.

38. Ibid., 70.

39. Thompson, "Artifacts: Travis Bean," 141.

40. Ibid.

41. Moseley, "Travis Bean Interview."

42. Art Thompson set the number of guitars and basses Travis Bean produced during 1974–1979 at between 1,700 and 3,650. Some of the variance is due to Travis Bean making surplus necks that other builders purchased and used. Thompson, "Artifacts: Travis Bean," 140–141.

43. Menn, "What's in a Name," 152.

44. Ken Achard, *The History and Development of the American Guitar* (Westport, CT: Bold Strummer, 1990), 156.

45. Thompson, "Artifacts: Travis Bean," 141.

46. André Millard (after the collaborative paper with Rebecca McSwain), "Heavy Metal: From Guitar Heroes to Guitar Gods," in Millard, ed., *The Electric Guitar: A History of an American Icon* (Baltimore: Johns Hopkins University Press, 2004), 175.

47. http://www.vintagekramer.com/alum.htm (accessed 31 March 2015).

48. Donald Brosnac, *An Introduction to Scientific Guitar Design* (1978; rptd., Westport, CT: Bold Strummer, 1996), 30.

49. Eric Clough, "Body for an Electronic Stringed Instrument," US Patent 4915004 A, filed 5 July 1988. http://www.google.com/patents/US4915004 (accessed 3 October 2014).

50. Charles Saufley, "Innovative Luthier Travis Bean Dies at 63," *Premier Guitar*, 11 July 2011. http://www.premierguitar.com/articles/Innovative_Luthier_Travis_Bean_Dies_at_63.

51. David Rothschild, "Schneller's Fine Specimens Strike a Chord," *Chicago Tribune*, 22 October 1993, 6. http://articles.chicagotribune.com/1993-10-22/entertainment/9310220332_1_yo-la-tengo-shrimp-boat-cmj-music-marathon.

52. Sean Guthrie, "O'Malley Sidesteps Metal Tag to Focus on Attitude," *Glasgow Herald*, 2 April 2014, 17.

53. Pat Gorman, *Instrument* (San Francisco: Chronicle Books, 2011), 76.

54. Kris Wernowsky, "Pensacola Guitar Maker Gains Reputation for Quality," *Pensacola News Journal*, 20 March 2011. http://www.theledger.com/article/20110320/NEWS/10 3205041?template=printart.

55. Interview with Tim Midyett by author, 22 March 2013.

56. Interview with Tim Becker by author, 25 February 2013.

57. Interview with Jodi Shapiro by author, 22 February 2013.

58. Ibid.

59. Martin, "Travis Bean," D8; Valerie J. Nelson, "Craftsman Revolutionized Sound, Life of Electric Guitars," *Washington Post*, 18 July 2011, B4.

60. Alan Greenwood and Gil Hembree, *The Official Vintage Guitar Magazine Price Guide 2007* (Bismarck, ND: Vintage Guitar, 2006), 240, 297; Alan Greenwood and Gil Hembree, *The Official Vintage Guitar Magazine Price Guide 2014* (Bismarck, ND: Vintage Guitar, 2013), 307–308.

61. Greenwood and Hembree, *The Official Vintage Guitar Magazine Price Guide 2007*, 244; Greenwood and Hembree, *The Official Vintage Guitar Magazine Price Guide 2014*, 314.

62. Greenwood and Hembree, *The Official Vintage Guitar Magazine Price Guide 2007*, 242; Greenwood and Hembree, *The Official Vintage Guitar Magazine Price Guide 2014*, 311.

63. Greenwood and Hembree, *The Official Vintage Guitar Magazine Price Guide 2007*, 167, 283–284; Greenwood and Hembree, *The Official Vintage Guitar Magazine Price Guide 2014*, 206.

64. See http://www.travisbeanguitars.com; and http://www.metalnecks.com (both accessed 22 June 2015).

65. Interview with Steve Albini by author, 31 December 2014.

66. Steve Albini, post on "Electrical Guitar Company" thread, *Electrical Audio* forum, 13 February 2014. http://www.electricalaudio.com/phpBB/viewtopic.php?f=4&t=13013& sid=3340778e42c4723c6b1e35eab8e70b06&start=3700#p1763316.

67. Interview with Steve Albini by author, 31 December 2014.

68. "Open Letter from Travis Bean." http://www.travisbeandesigns.com/files/docu ments/Open-Letter-from-Travis-Bean.pdf (accessed 7 May 2015).

69. "Obstructures: People," http://obstructures.org/people; "ABC Aluminum Guitar Prototype 1999," http://obstructures.org/work/thing/17/abc-aluminum-guitar (both accessed 7 July 2014).

70. Kevin Burkett, post on "Travis Bean Owners!" thread, *Electrical Audio* forum, 1 November 2003. http://www.electricalaudio.com/phpBB3/viewtopic.php?f=4&t=375&p =2671#p3687.

71. Wernowsky, "Pensacola Guitar Maker."

72. Interview with Kevin Burkett by author, 25 June 2013.

73. "TB500," Electrical Guitar Company. http://www.electricalguitarcompany.com /models (accessed 14 December 2014).

74. Interview with Kevin Burkett by author, 25 June 2013.

75. Ibid.

76. Wernowsky, "Pensacola Guitar Maker."

77. Interview with Tim Midyett by author, 30 May 2013.

78. Brad Angle, "Dear Guitar Hero: Buzz Osborne of the Melvins Discusses Aluminum Guitars, Tool's Adam Jones and Growing Up with Kurt Cobain," *Guitar World*,

4 September 2013. http://www.guitarworld.com/dear-guitar-hero-buzz-osborne-melvins-discusses-aluminum-guitars-tools-adam-jones-and-growing-kurt-cobain.

79. Ibid.

80. Interview with Chris Rasmussen by author, 24 February 2013.

81. Wernowsky, "Pensacola Guitar Maker."

82. Interview with Tim Midyett by author, 30 May 2013.

83. Interview with Kevin Burkett by author, 25 June 2013.

84. Interview with Conan Neutron by author, 20 June 2015.

85. Ibid.

86. Christopher Borrelli, "The Ax Collector," *Chicago Tribune*, 16 August 2012, 4–1, 4–6, 4–7.

87. Interview with Conan Neutron by author, 20 June 2015.

88. Martin, "Travis Bean," D8; http://www.travisbeandesigns.com (accessed 17 February 2015).

CONCLUSION: **Designing for Sustainability**

1. William McDonough and Michael Braungart, *The Upcycle: Beyond Sustainability, Designing for Abundance* (New York: North Point, 2013), 6.

2. Adam Minter, *Junkyard Planet: Travels in the Billion-Dollar Trash Trade* (New York: Bloomsbury, 2013), 256.

3. Samantha MacBride, *Recycling Reconsidered: The Present Failure and Future Promise of Environmental Action in the United States* (Cambridge, MA: MIT Press, 2011), 110–112.

4. US Geological Survey, "Aluminum Statistics," in T. D. Kelly and G. R. Matos, comps., *Historical Statistics for Mineral and Material Commodities in the United States: U.S. Geological Survey Data*, ser. 140. http://minerals.usgs.gov/minerals/pubs/historical-statistics/ds140-alumi.xlsx (accessed 21 December 2015).

5. Jennifer Gitlitz, *Trashed Cans: The Global Environmental Impacts of Aluminum Can Wasting in America* (Arlington, VA: Container Recycling Institute, 2002), 17.

6. Mimi Sheller, *Aluminum Dreams: The Making of Light Modernity* (Cambridge, MA: MIT Press, 2014), 180, 195, 197.

7. Andy Home, "China, the Aluminum Giant That's Still Growing," *Reuters*, 20 July 2015. http://www.reuters.com/article/us-aluminium-production-ahome-idUSKCN0PU21C20150720; Biman Mukherji, "Aluminum Rises as Chinese Producers Vow to Cut Production," *Wall Street Journal*, 11 December 2015. http://www.wsj.com/articles/aluminum-rises-as-chinese-producers-vow-to-cut-production-1449833905.

8. An excellent discussion of the problems establishing a scale for the environmental effects of plastics is Max Liboiron, "Redefining Pollution: Plastics in the Wild," PhD diss., New York University, 2012.

9. Philip White, Louise St. Pierre, and Steve Belletire, *Okala Practitioner: Integrating Ecological Design* (Phoenix, AZ: Okala Team, 2013), chap. 4, 4.

10. Thorstein Veblen, *The Theory of the Leisure Class: An Economic Study of Institutions* (New York: Macmillan, 1899); Juliet B. Schor, *The Overspent American: Why We Want What We Don't Need* (New York: Basic, 1997).

Index

Page numbers in italics signify photographs.

About the Author

Carl A. Zimring is an associate professor of sustainability studies at the Pratt Institute. He is the author of *Cash for Your Trash: Scrap Recycling in America* (2005) and *Clean and White: A History of Environmental Racism in the United States* (2015) and the co-editor (with William L. Rathje) of the *Encyclopedia of Consumption and Waste: The Social Science of Garbage* (2012).